THE ROMANCE OF MODERN MECHANISM

A MECHANICAL SCULPTOR

The lower illustration shows the Wenzel Sculpturing Machine at work on two blocks of stone ranged one on each side of a model. This machine can make four copies simultaneously from one original. The upper illustration shows the quality of work done by the automatic sculptor.

THE ROMANCE OF MODERN MECHANISM

WITH INTERESTING DESCRIPTIONS IN NON-TECHNICAL LANGUAGE OF WONDERFUL MACHINERY AND MECHANICAL DEVICES AND MARVELLOUSLY DELICATE SCIENTIFIC INSTRUMENTS, &c., &c.

BY

ARCHIBALD WILLIAMS,
B.A., Oxon., F.R.G.S.

AUTHOR OF
"THE ROMANCE OF MODERN INVENTION," "THE ROMANCE OF MODERN MINING," "THE ROMANCE OF MODERN ENGINEERING," "THE ROMANCE OF MODERN EXPLORATION,"
&c. &c.

WITH THIRTY ILLUSTRATIONS

LONDON
SEELEY AND CO. LIMITED
38 GREAT RUSSELL STREET

1910

UNIFORM WITH THIS VOLUME

THE LIBRARY OF ROMANCE

Extra Crown 8vo. With many illustrations. 5s. each

"Splendid volumes."—*The Outlook.*

"This series has now won a considerable and well deserved reputation."—*The Guardian.*

"Each volume treats its allotted theme with accuracy, but at the same time with a charm that will commend itself to readers of all ages. The root idea is excellent, and it is excellently carried out, with full illustrations and very prettily designed covers."—*The Daily Telegraph.*

By Prof. G. F. SCOTT ELLIOT, M.A., B.Sc.
THE ROMANCE OF SAVAGE LIFE
THE ROMANCE OF PLANT LIFE
THE ROMANCE OF EARLY BRITISH LIFE

By EDWARD GILLIAT, M.A.
THE ROMANCE OF MODERN SIEGES

By JOHN LEA, M.A.
THE ROMANCE OF BIRD LIFE

By JOHN LEA, M.A., & H. COUPIN, D.Sc.
THE ROMANCE OF ANIMAL ARTS AND CRAFTS

By SIDNEY WRIGHT
THE ROMANCE OF THE WORLD'S FISHERIES

By the Rev. J. C. LAMBERT, M.A., D.D.
THE ROMANCE OF MISSIONARY HEROISM

By G. FIRTH SCOTT
THE ROMANCE OF POLAR EXPLORATION

By ARCHIBALD WILLIAMS, B.A. (Oxon.), F.R.G.S.
THE ROMANCE OF EARLY EXPLORATION
THE ROMANCE OF MODERN EXPLORATION
THE ROMANCE OF MODERN MECHANISM
THE ROMANCE OF MODERN INVENTION
THE ROMANCE OF MODERN ENGINEERING
THE ROMANCE OF MODERN LOCOMOTION
THE ROMANCE OF MODERN MINING

By CHARLES R. GIBSON, A.I.E.E.
THE ROMANCE OF MODERN PHOTOGRAPHY

THE ROMANCE OF MODERN ELECTRICITY
THE ROMANCE OF MODERN MANUFACTURE

By EDMUND SELOUS
THE ROMANCE OF THE ANIMAL WORLD
THE ROMANCE OF INSECT LIFE

By AGNES GIBERNE
THE ROMANCE OF THE MIGHTY DEEP

By E. S. GREW, M.A.
THE ROMANCE OF MODERN GEOLOGY

By J. C. PHILIP, D.Sc., Ph.D.
THE ROMANCE OF MODERN CHEMISTRY

SEELEY & CO., LIMITED

INTRODUCTION

In the beginning a man depended for his subsistence entirely upon his own efforts, or upon those of his immediate relations and friends. Life was very simple in those days: luxury being unknown, and necessity the factor which guided man's actions at every turn. With infinite labour he ground a flint till it assumed the shape of a rough arrow-head, to be attached to a reed and shot into the heart of some wild beast as soon as he had approached close enough to be certain of his quarry. The meat thus obtained he seasoned with such roots and herbs as nature provided—a poor and scanty choice. Presently he discovered that certain grains supported life much better than roots, and he became an agriculturist. But the grain must be ground; so he invented a simple mill—a small stone worked by hand over a large one; and when this method proved too tedious he so shaped the stones' surfaces that they touched at all points, and added handles by which the upper stone could be revolved.

With the discovery of bronze, and, many centuries later, of iron, his workshop equipment rapidly improved. He became an expert boat- and house-builder, and multiplied weapons of offence and defence. Gradually separate crafts arose. One man no longer depended on his individual efforts, but was content to barter his own work for the products of another man's labour, because it became evident that specialisation promoted excellence of manufacture.

A second great step in advance was the employment of machinery, which, when once fashioned by hand, saved an enormous amount of time and trouble—the pump, the blowing bellows, the spinning-wheel, the loom. But all had to be operated by human effort, sometimes replaced by animal power.

With the advent of the steam-engine all industry bounded forward again. First harnessed by Watt, Giant Steam has become a commercial and political power. Everywhere, in mill and factory, locomotive, ship, it has increased the products which lend ease and comfort to modern life; it is the great ally of invention, and the ultimate agent for transporting men and material from one point on the earth's surface to another.

Try as we may, we cannot escape from our environment of mechanism, unless we are content to revert to the loincloth and spear of the savage. Society has become so complicated that the utmost efforts of an individual are, after all, confined to a very narrow groove. The days of the Jack-of-all-trades are over. Success in life, even bare subsistence, depends on the concentration of one's faculties upon a very limited daily routine. "Let the cobbler stick to his last" is a maxim which carries an ever-increasing force.

The better to realise how dependent we are on the mechanisms controlled by the thousand and one classes of workmen, let us consider the surroundings, possessions, and movements of the average, well-to-do business man.

At seven o'clock he wakes, and instinctively feels beneath his pillow for his watch, a most marvellous assemblage of delicate parts shaped by wonderful machinery. Before stepping into his bath he must turn a tap, itself a triumph of mechanical skill. The razor he shaves with, the mirror which helps him in the operation, the very brush and soap, all are machine-made. With his clothes he adds to the burden of his indebtedness to mechanism. The power-loom span the linen for his shirts, the cloth for his outer garments. Shirts and collars are glossy from the treatment of the steam laundry, where machinery is rampant. His boots, kept shapely by machine-made lasts, should remind him that mechanical devices have played a large part in their manufacture, very possibly the human hand has scarcely had a single duty to perform.

He goes downstairs, and presses an electric button. Mechanism again. While waiting for his breakfast his eye roves carelessly over the knives, spoons, forks, table, tablecloth, wall-paper, engravings, carpet, cruet-stand—all machine-made in a larger or less degree. The very coals blazing in the grate were won by machinery; the marble of the mantelpiece was shaped and polished by machinery; also the fire-irons, the chairs, the hissing kettle. Machinery stares at him from the loaf on its machine-made board. Machines prepared the land, sowed, harvested, threshed, ground, and probably otherwise prepared the grain for baking. Machines ground his salt, his coffee. Machinery aided the capture of the tempting sole; helped to cure the rasher of bacon; shaped the dishes, the plates, the coffee-pot.

Whirr-r-r! The motor-car is at the door, throbbing with the impulses of its concealed machinery. Our friend therefore puts on his machine-made gloves and hat and sallies forth. That wonderful motor, the product of the most up-to-date, scientific, and mechanical appliances, bears him swiftly over roads paved with machine-crushed stone and flattened out by a steam-roller. A book might be reserved to the motor alone; but we must refrain, for a few minutes' travel has brought the horseless carriage to the railway station. Mr. Smith, being the holder of a season ticket, does not trouble the clerk who is stamping pasteboards with a most ingenious contrivance for automatically impressing dates and numbers on them. He strolls out on the platform and buys the morning paper, which, a few hours before, was being battered about by one of the most wonderful machines that ever was devised by the brain of man. Mr. Smith doesn't bother his head with thoughts of the printing-press. Its products are all round him, in timetables and advertisements. Nor does he ponder upon the giant machinery which crushed steel ingots into the gleaming rails that stretch into the far distance; nor upon the marvellous interlocking mechanism of the signal-box at the platform-end; nor upon the electric wires thrumming overhead. No! he had seen all these things a thousand times before, and probably feels little of the romance which lies so thickly upon them.

A whistle blows. The "local" is approaching, with its majestic locomotive—a very orgy of mechanism—its automatic brakes, its thousand parts all shaped by mechanical devices,—steam saws, planes, lathes, drills, hammers, presses. In obedience to a little lever the huge mass comes quickly to rest; the steam pump on the engine commences to gasp; a minute later another lever moves, and Mr. Smith is fairly on his way to business.

Arrived at the metropolis, he presses electricity into his service, either on an electric tram or on a subterranean train. In the latter case he uses an electric lift, which lowers him into the bowels of the earth, to pass him on to the current-propelled cars, driven by power generated in far-away stations.

His office is stamped all over with the seal of mechanism. In the lobby are girls hammering on marvellous typewriters; on his desk rests a telephone, connected through wires and most elaborately equipped exchanges with all parts of the country. To get at his private and valuable papers Mr. Smith must have recourse to his bunch of keys, which, with their corresponding locks, represent ingenuity of a high degree. All day long he is in the grasp of mechanism; not even at lunch time can he escape it, for the food set before him at the restaurant has been cooked by the aid of special kitchen machinery.

And when the evening draws on Mr. Smith touches a switch to turn his darkness into light, wrung through many wonderful processes from the stored illumination of coal.

Were we to trace the daily round of the clerk, artisan, scientist, engineer, or manufacturer, we should be brought into contact with a thousand other mechanical appliances. Space forbids such a tour of inspection; but in the following pages we may rove here and there through the workshops of the world, gleaning what seems to be of special interest to the general public, and weaving round it, with a machine-made pen, some of the romance which is apt to be lost sight of by the most marvellous of all creations—Man.

AUTHOR'S NOTE

THE author desires to express his indebtedness to the following gentlemen for the kind help they have afforded him in connection with the gathering of materials for the letterpress and illustration of this book:—

The proprietors of *Cassier's Magazine*, *The Magazine of Commence*, *The World's Work*, *The Motor Boat*; The Rexer Automatic Machine Gun Co.; The Diesel Oil Engine Co.; The Cambridge Scientific Instrument Co.; The Marconi Wireless Telegraphy Co.; The Temperley Transporter Co.; Messrs. de Dion, Bouton and Co.; Messrs. Merryweather and Sons; Mr. A. Crosby Lockwood; Mr. Dan Albone; Mr. J. B. Diplock; Mr. W. H. Oatway; The National Cash Register Co.; The Wenzel Sculpturing Machine Co.; Mr. E. W. Gaz; Sir W. G. Armstrong, Whitworth and Co.; The International Harvester Co. and Messrs. Gwynne and Co.

TABLE OF CONTENTS

	PAGE
INTRODUCTION	v
AUTHOR'S NOTE	xi

CHAPTER I

DELICATE INSTRUMENTS — WATCHES AND CHRONOMETERS — THE MICROTOME — THE DIVIDING ENGINE — MEASURING MACHINES	22

CHAPTER II

CALCULATING MACHINES	44

CHAPTER III

WORKSHOP MACHINERY — THE LATHE — PLANING MACHINES — THE STEAM HAMMER — HYDRAULIC TOOLS — ELECTRICAL TOOLS IN THE SHIPYARD	60

CHAPTER IV

PORTABLE TOOLS	87

CHAPTER V

THE PEDRAIL: A WALKING STEAM-ENGINE	93

CHAPTER VI

INTERNAL COMBUSTION ENGINES —
OIL ENGINES — ENGINES WORKED
WITH PRODUCER GAS — BLAST
FURNACE GAS ENGINES   106

CHAPTER VII

MOTOR-CARS — THE MOTOR OMNIBUS
— RAILWAY MOTOR-CARS   121

CHAPTER VIII

THE MOTOR AFLOAT — PLEASURE
BOATS — MOTOR LIFEBOATS —
MOTOR FISHING BOATS — A MOTOR
FIRE FLOAT — THE MECHANISM OF
THE MOTOR BOAT — THE TWO-
STROKE MOTOR — MOTOR BOATS FOR
THE NAVY   137

CHAPTER IX

THE MOTOR CYCLE   158

CHAPTER X

FIRE ENGINES   167

CHAPTER XI

FIRE-ALARMS AND AUTOMATIC FIRE
EXTINGUISHERS   173

CHAPTER XII

THE MACHINERY OF A SHIP — THE
REVERSING ENGINE — MARINE
ENGINE SPEED GOVERNORS — THE
STEERING ENGINE — BLOWING AND

VENTILATING APPARATUS — PUMPS — FEED HEATERS — FEED-WATER FILTERS — DISTILLERS — REFRIGERATORS — THE SEARCH-LIGHT — WIRELESS TELEGRAPHY INSTRUMENTS — SAFETY DEVICES — THE TRANSMISSION OF POWER ON A SHIP 183

CHAPTER XIII

"THE NURSE OF THE NAVY" 212

CHAPTER XIV

THE MECHANISM OF DIVING 216

CHAPTER XV

APPARATUS FOR RAISING SUNKEN SHIPS AND TREASURE 225

CHAPTER XVI

THE HANDLING OF GRAIN — THE ELEVATOR — THE SUCTION PNEUMATIC GRAIN-LIFTER — THE PNEUMATIC BLAST GRAIN-LIFTER — THE COMBINED SYSTEM 226

CHAPTER XVII

MECHANICAL TRANSPORTERS AND CONVEYERS — ROPEWAYS — CABLEWAYS — TELPHERAGE — COALING WARSHIPS AT SEA 235

CHAPTER XVIII

AUTOMATIC WEIGHERS	249

CHAPTER XIX

TRANSPORTER BRIDGES	253

CHAPTER XX

BOAT- AND SHIP-RAISING LIFTS	259

CHAPTER XXI

A SELF-MOVING STAIRCASE	270

CHAPTER XXII

PNEUMATIC MAIL TUBES	276

CHAPTER XXIII

AN ELECTRIC POSTAL SYSTEM	288

CHAPTER XXIV

AGRICULTURAL MACHINERY — PLOUGHS — DRILLS AND SEEDERS — REAPING MACHINES — THRESHING MACHINES — PETROL-DRIVEN FIELD MACHINERY — ELECTRICAL FARMING MACHINERY	292

CHAPTER XXV

DAIRY MACHINERY — MILKING MACHINES — CREAM SEPARATORS — A MACHINE FOR DRYING MILK	304

CHAPTER XXVI

SCULPTURING MACHINES 309

CHAPTER XXVII

AN AUTOMATIC RIFLE — A BALL-
BEARING RIFLE 319

THE ROMANCE OF MODERN MECHANISM

CHAPTER I

DELICATE INSTRUMENTS

WATCHES AND CHRONOMETERS — THE MICROTOME — THE DIVIDING ENGINE — MEASURING MACHINES

Owing to the universal use of watches, resulting from their cheapness, the possessor of a pocket timepiece soon ceases to take a pride in the delicate mechanism which at first added an inch or two to his stature. At night it is wound up mechanically, and thrust under the pillow, to be safe from imaginary burglars and handy when the morning comes. The awakened sleeper feels small gratitude to his faithful little servant, which all night long has been beating out the seconds so that its master may know just where he is with regard to "the enemy" on the morrow. At last a hand is slipped under the feather-bag, and the watch is dragged from its snug hiding-place. "Bother it," says the sleepy owner, "half-past eight; ought to have been up an hour ago!" and out he tumbles. Dressing concluded, the watch passes to its day quarters in a darksome waistcoat pocket, to be hauled out many times for its opinion to be taken.

The real usefulness of a watch is best learnt by being without one for a day or two. There are plenty of clocks about, but not always in sight; and one gradually experiences a mild irritation at having to step round the corner to find out what the hands are doing.

A truly wonderful piece of machinery is a watch—even a cheap one. An expensive, high-class article is worthy of our admiration and respect. Here is one that has been in constant use for fifty years. Twice a second its little balance-wheel revolves on its jewelled bearings. Allowing a few days for repairs, we find by calculation that the watch has made no less than three thousand million movements in the half-century! And still it goes ticking on, ready to do another fifty years' work. How beautifully tempered must be the springs and the steel faces which are constantly rubbing against jewel or metal! How perfectly cut the teeth which have engaged one another times innumerable without showing appreciable wear!

The chief value of a good watch lies in its accuracy as a time-keeper. It is, of course, easy to correct it by standard clocks in the railway stations or public buildings; but one may forget to do this, and in a week or two a loss of a few minutes may lead to one missing a train, or being late for an important engagement. Happy, therefore, is the man who, having set his watch to "London time," can rely on its not varying from accuracy a minute in a week—a feat achieved by many watches.

The old-fashioned watch was a bulky affair, protected by an outer case of ample proportions. From year to year the size has gradually diminished, until we can now purchase a reliable article no thicker than a five-shilling piece, which will not offend the most fastidious dandy by disarranging the fit of his clothes. Into the space of a small fraction of an inch is crowded all the usual mechanism, reduced to the utmost fineness. Watches have even been constructed small enough to form part of a ring or earring, without losing their time-keeping properties.

For practical purposes, however, it is advantageous to have a timepiece of as large a size as may be convenient, since the difficulties of adjustment and repair increase with decreasing proportions. The ship's chronometer, therefore, though of watch construction, is a big affair as compared with the pocket timepiece; for above all things it must be accurate.

The need for this arises from the fact that nautical reckonings made by the observation of the heavenly bodies include an element of *time*. We will suppose a vessel to be at sea out of sight of land. The captain, by referring to the dial of the "mechanical log," towed astern, can reckon pretty accurately how *far* the vessel has travelled since it left port; but owing to winds and currents he is not certain of the position on the globe's surface at which his ship has arrived. To locate this exactly he must learn (*a*) his longitude, *i.e.* distance E. or W. of Greenwich, (*b*) his latitude, *i.e.* distance N. or S. of the Equator. Therefore, when noon approaches, his chronometers and sextant are got out, and at the moment when the sun crosses the meridian the time is taken. If this moment happens to coincide with four o'clock on the chronometers he is as far west of Greenwich as is represented by four twenty-fourths of the 360° into which the earth's circumference is divided; that is, he is in longitude 60° W. The sextant gives him the angle made by a line drawn to the sun with another drawn to the horizon, and from that he calculates his latitude. Then he adjourns to the chart-room, where, by finding the point at which the lines of longitude and latitude intersect, he establishes his exact position also.

When the ship leaves England the chronometer is set by Greenwich time, and is never touched afterwards except to be wound once a day. In order that any error may be reduced to a minimum a merchant ship carries at least two chronometers, a man-of-war at least three, and a surveying vessel as many as a dozen. The average reading of the chronometers is taken to work by.

Taking the case of a single chronometer, it has often to be relied on for months at a time, and during that period has probably to encounter many changes of temperature. If it gains or loses from day to day, and that *consistently*, it may still be accounted reliable, as the amount of error will be allowed for in all calculations. But should it gain one day and lose another, the accumulated errors would, on a voyage of several months, become so considerable as to imperil seriously the safety of the vessel if navigating dangerous waters.

As long ago as 1714 the English Government recognised the importance of a really reliable chronometer, and in that year passed an Act offering rewards of £10,000, £15,000, and £20,000 to anybody who should produce a chronometer that would fix longitude within sixty, forty, and thirty miles respectively of accuracy. John Harrison, the son of a Yorkshire carpenter, who had already invented the ingenious "gridiron pendulum" for compensating clocks, took up the challenge. By 1761 he had made a chronometer of so perfect a nature that during a voyage to Jamaica that year, and back the next, it lost only 1 min. $54\frac{1}{2}$ sec. As this would enable a captain to find his longitude within eighteen miles in the latitude of Greenwich, Harrison claimed, and ultimately received, the maximum reward.

It was not till nearly a century later that Thomas Earnshaw produced the "compensation balance," now generally used on chronometers and high-class watches. In cheap watches the balance is usually a little three-spoked wheel, which at every tick revolves part of a turn and then flies back again. This will not suffice for very accurate work, because the "moment of inertia" varies at different temperatures. To explain this term let us suppose that a man has a pound of metal to make into a wheel. If the wheel be of small diameter, you will be able to turn it first one way and then the other on its axle quite easily. But should it be melted down and remade into a wheel of four times the diameter, with the same amount of metal as before in the rim, the difficulty of suddenly reversing its motion will be much increased. The weight is the same, but the speed of the rim, and consequently its momentum, is greater. It is evident from this that, if a wheel of certain size be driven by a spring of constant strength, its oscillations will be equal in time; but if a rise of temperature should lengthen the spokes the speed would fall, because the spring would have more work to do; and, conversely, with a fall of temperature the speed would rise. Earnshaw's problem was to construct a balance wheel that should be able to keep its "moment of inertia" constant under all circumstances. He therefore used only two spokes to his wheel, and to the outer extremity of each attached an almost complete semicircle of rim, one end being attached to the spoke, the other all but meeting the other spoke. The rim-pieces were built up of an outer strip of brass, and an inner strip of steel welded together. Brass expands more rapidly than steel, with the result that a bar compounded of these two metals would, when heated, bend towards the hollow side. To the rim-pieces were attached sliding weights, adjustable to the position found by experiment to give the best results.

We can now follow the action of the balance wheel. It runs perfectly correctly at, say, a temperature of 60°. Hold it over a candle. The spokes lengthen, and carry the rim-pieces *outwards* at their fixed ends; but, as the pieces themselves bend inwards at their free ends, the balance is restored. If the balance were placed in a refrigerating machine, the spokes would shorten, but the rim-pieces would bend outwards.

As a matter of fact, the "moment of inertia" cannot be kept quite constant by this method, because the variation of expansion is more rapid in cold than in heat; so that, though a balance might be quite reliable between 60° and 100°, it would fail between 30° and 60°. So the makers fit their balances with what is called a *secondary* compensation, the effect of which is to act more quickly in high than in low temperatures. This could not well be explained without diagrams, so a mere mention must suffice.

Another detail of chronometer making which requires very careful treatment is the method of transmitting power from the main spring to the works. As the spring uncoils, its power must decrease, and this loss must be counterbalanced somehow. This is managed by using the "drum and fusee" action, which may be seen in some clocks and in many old watches. The drum is cylindrical, and contains the spring. The fusee is a tapering shaft, in which a spiral groove has been cut from end to end. A very fine chain connects the two parts. The key is applied to the fusee, and the chain is wound off the drum on to the larger end of the fusee first. By the time that the spring has been fully wound, the chain has reached the fusee's smaller extremity. If the fusee has been turned to the correct taper, the driving power of the spring will remain constant as it unwinds, for it gets least leverage over the fusee when it is strongest, and most when it is weakest, the intermediate stages being properly proportioned. To test this, a weighted lever is attached to the key spindle, with the weight so adjusted that the fully wound spring has just sufficient power to lift it over the topmost point of a revolution. It is then allowed a second turn, but if the weight now proves excessive something must be wrong, and the fusee needs its diameter reducing at that point. So the test goes on from turn to turn, and alterations are made until every revolution is managed with exactly the same ease.

The complete chronometer is sent to Greenwich observatory to be tested against the Standard Clock, which, at 10 a.m., flashes the hour to other clocks all over Great Britain. In a special room set apart for the purpose are hundreds of instruments, some hanging up, others lying flat. Assistants make their rounds, noting the errors on each. The temperature test is then applied in special ovens, and finally the article goes back to the maker with a certificate setting forth its performances under different conditions. If the error has been consistent the instrument is sold, the buyer being informed exactly what to allow for each day's error. At the end of the voyage he brings his chronometer to be tested again, and, if necessary, put right.

Here are the actual variations of a chronometer during a nineteen-day test, before being used:—

Day.	Gain in tenths of seconds.	Day.	Gain in tenths of seconds.
1st	½	11th	4
2nd	3	12th	3
3rd	4	13th	3
4th	4	14th	4
5th	½	15th	5
6th	3	16th	2
7th	0	17th	3
8th	0	18th	5
9th	4½	19th	1
10th	3		

An average gain of just over one quarter of a second per diem! Quite extraordinary feats of time-keeping have been recorded of chronometers on long voyages. Thus a chronometer which had been to Australia *viâ* the Cape and back *viâ* the Red Sea was only fifteen seconds "out"; and the *Encyclopædia Britannica* quotes the performance of the three instruments of s.s. *Orellana*, which between them accumulated an error of but 2·3 seconds during a sixty-three-day trip.

An instrument which will cut a blood corpuscle into several parts—that's the Microtome, the "small-cutter," as the name implies.

For the examination of animal tissues it is necessary that they should be sliced very fine before they are subjected to the microscope. Perhaps a tiny muscle is being investigated and cross sections of it are needed. Well, one cannot pick up the muscle and cut slices off it as you would off a German sausage. To begin with, it is difficult even to pick the object up; and even if pieces one-hundredth of an inch long were detached they would still be far too large for examination.

So, as is usually the case when our unaided powers prove unequal to a task, we have recourse to a machine. There are several types of microtomes, each preferable for certain purposes. But as in ordinary laboratory work the Cambridge Rocking Microtome is used, let us give our special attention to this particular instrument. It is mounted on a strong cast-iron bed, a foot or so in length and four to five inches wide. Towards one end rise a couple of supports terminating in knife-edges, which carry a cross-bar, itself provided with knife-edges top and bottom, those on the top supporting a second transverse bar. Both bars have a long leg at right angles, giving them the appearance of two large T's superimposed one on the other; but the top T is converted into a cross by a fourth member—a sliding tube which projects forward towards a frame in which is clamped a razor, edge upwards.

The tail of the lower T terminates in a circular disc, pierced with a hole to accommodate the end of a vertical screw, which has a large circular head with milled edges. The upper T is rocked up and down by a cord and spring, the handle actuating the cord also shifting on the milled screw-head a very small distance every time it is rocked backwards and forwards. As the screw turns, it gradually raises the tail of the lower member, and by giving its cross-bar a tilt brings the tube of the upper member appreciably nearer the razor. The amount of twist given to the screw at each stroke can be easily regulated by a small catch.

When the microscopist wishes to cut sections he first mounts his object in a lump of hard paraffin wax, coated with softer wax. The whole is stuck on to the face of the tube, so as to be just clear of the razor.

The operator then seizes the handle and works it rapidly until the first slice is detached by the razor. Successive slices are stuck together by their soft edges so as to form a continuous ribbon of wax, which can be picked up easily and laid on a glass slide. The slide is then warmed to melt the paraffin, which is dissolved away by alcohol, leaving the atoms of tissue untouched. These, after being stained with some suitable medium, are ready for the microscope.

A skilful user can, under favourable conditions, cut slices *one twenty-five thousandth* of an inch thick. To gather some idea of what this means we will imagine that a cucumber one foot long and one and a-half inches in diameter is passed through this wonderful guillotine. It would require no less than 700 dinner-plates nine inches across to spread the pieces on! If the slices were one-eighth of an inch thick, the cucumber, to keep a proportionate total size, would be 260 feet long. After considering these figures we shall lose some of the respect we hitherto felt for the men who cut the ham to put inside luncheon-bar sandwiches.

In the preceding pages frequent reference has been made to index screws, exactly graduated to a convenient number of divisions. When such screws have to be manufactured in quantities it would be far too expensive a matter to measure each one separately. Therefore machinery, itself very carefully graduated, is used to enable a workman to transfer measurements to a disc of metal.

If the index-circle of an astronomical telescope—to take an instance—has to be divided, it is centred on a large horizontal disc, the circumference of which has been indented with a large number of teeth. A worm-screw engages these teeth tangentially (*i.e.* at right angles to a line drawn from the centre of the plate to the point of engagement). On the shaft of the screw is a ratchet pinion, in principle the same as the bicycle free-wheel, which, when turned one way, also twists the screw, but has no effect on it when turned the other way. Stops are put on the screw, so that it shall rotate the large disc only the distance required between any two graduations. The divisions are scribed on the index-circle by a knife attached to a carriage over and parallel to the disc. The DIVIDING ENGINE used for the graduation of certain astronomical instruments probably constitutes the most perfect machine ever made. In an address to the Institution of Mechanical Engineers,[1] the President, Mr. William Henry Maw, used the following words: "The most recently constructed machine of the kind of which I am aware—namely, one made by Messrs. Warner and Swasey, of Cleveland, U.S.A.—is capable of automatically cutting the graduations of a circle with an error in position not exceeding one second of arc. (A second of an arc is approximately the angle subtended by a halfpenny at a distance of three miles.) This means that on a 20-inch circle the error in position of any one graduation shall not exceed $\frac{1}{20,000}$ inch. Now, the finest line which would be of any service for reading purposes on such a circle would probably have a width equal to quite ten seconds of arc; and it follows that the minute V-shaped cut forming this line must be so absolutely symmetrical with its centre line throughout its length, that the position of this centre may be determined within the limit of error just stated by observations of its edges, made by aid of the reading micrometer and microscope. I may say that after the machine just mentioned had been made, it took *over a year's hard work* to reduce the maximum error in its graduations from one and a-half to one second of arc."

The same address contains a reference to the great Yerkes telescope, which though irrelevant to our present chapter, affords so interesting an example of modern mechanical perfection that it deserves parenthetic mention.

The diameter of a star of the seventh magnitude as it appears in the focus of this huge telescope is $\frac{1}{2,500}$ inch. The spiders' webs stretched across the object glass are about $\frac{1}{6,000}$ inch in diameter. "The problem thus is," says Mr. Maw, "to move this twenty-two ton mass (the telescope) with such steadiness in opposition to the motion of the earth, that a star disc $\frac{1}{2,500}$ inch in diameter can be kept threaded, as it were, upon a spider's web $\frac{1}{6,000}$ inch in diameter, carried at a radius of thirty-two feet from the centre of motion. I think that you will agree that this is a problem in mechanical engineering demanding no slight skill to solve; but it has been solved, and with the most satisfactory results." The motions are controlled electrically; and respecting them Professor Barnard, one of the chief observers with this telescope, some time ago wrote as follows: "It is astonishing to see with what perfect instantaneousness the clock takes up the tube. The electric slow motions are controlled from the eye end. So exact are they that a star can be brought from the edge of the field and stopped instantaneously behind the micrometer wire."

Dividing engines are used for ruling parallel lines on glass and metal, to aid in the measurements of microscopical objects or the wave-lengths of light. A *diffraction grating*, used for measuring the latter, has the lines so close together that they would be visible only under a powerful microscope. Glass being too brittle, a special alloy of so-called *speculum* metal is fashioned into a highly polished plate, and this is placed in the machine. A delicate screw arrangement gradually feeds the plate forwards under the diamond point, which is automatically drawn across the plate between every two movements. Professor H. A. Rowlands has constructed a parallel dividing engine which has ruled as many as 120,000 lines to the inch. To get a conception of these figures we must once again resort to comparison. Let us therefore take a furrow as a line, and imagine a ploughman going up and down a field 120,000 times. If each furrow be eight inches wide, the field would require a breadth of nearly *fourteen miles* to accommodate all the furrows! Again, supposing that a plate six inches square were being ruled, the lines placed end to end would extend for seventy miles!

Professor Rowlands' machine does the finest work of this kind. Another very perfect instrument has been built by Lord Blythswood, and as some particulars of it have been kindly supplied, they may fitly be appended.

If a first-class draughtsman were asked how many parallel

straight lines he would rule within the space of one inch, it is doubtful whether he would undertake more than 150 to 200 lines. Lord Blythswood's machine can rule fourteen parallel lines on a space equivalent to the *edge* of the finest tissue paper. So delicate are the movements of the machine that it must be protected from variations of temperature, which would contract or expand its parts; so the room in which it stands is kept at an even heat by automatic apparatus, and to make things doubly sure the engine is further sheltered in a large case having double walls inter-packed with cotton wool.

In constructing the machine it was found impossible, with the most scientific tools, to cut a toothed wheel sufficiently accurate to drive the mechanism, but the errors discovered by microscopes were made good by the invention of a small electro-plating brush, which added the thinnest imaginable layer of metal to any tooth found deficient.

During the process of ruling a grating of only a few square inches area, the machine must be left severely alone in its closed case. The slightest jar would cause unparallelism of a few lines, and the ruin of the whole grating. So for several days the diamond point has its own way, moving backwards and forwards unceasingly over the hard metal, in which it chases tiny grooves. At the end the plate has the appearance of mother-of-pearl, which is, in fact, one of nature's diffraction gratings, breaking up white light into the colours of the spectrum.

You will be able to understand that these mechanical gratings are expensive articles. Sometimes the diamond point breaks half-way through the ruling, and a week's work is spoilt. Also the creation of a reliable machine is a very tedious business. Ten pounds per square inch of grating is a low price to pay.

The greatest difficulty met with in the manufacture of the dividing engine is that of obtaining a mathematically correct screw. Turning on a lathe produces a very rough spiral, judged scientifically. Some threads will be deeper than others, and differently spaced. The screw must, therefore, be ground with emery and oil introduced between it and a long nut which is made in four segments, and provided with collars for tightening it up against the screw. Perhaps a fortnight may be expended over the grinding. Then the screw must undergo rigid tests, a nut must be made for it, and it has to be mounted in proper bearings. The explanation of the method of eliminating errors being very technical, it is omitted; but an idea of the care required may be gleaned from Professor Rowlands' statement that an uncorrected error of $\frac{1}{300,000}$ of an inch is quite sufficient to ruin a grating!

In the Houses of Parliament there is kept at an even temperature a bronze rod, thirty-eight inches long and an inch square in section. Near the ends are two wells, rather more than half an inch deep, and at the bottom of the wells are gold studs, each engraved with a delicate cross line on their polished surfaces. The distance between the lines is the imperial yard of thirty-six inches.

The bar was made in 1844 to replace the Standard destroyed in 1834, when both Houses of Parliament were burned. The original Standard was the work of Bird, who produced it in 1760. In June, 1824, an Act had been passed legalising this Standard. It says:—

"The same Straight Line or Distance between the Centers of the said Two Points in the said Gold Studs in the said Brass Rod, the Brass being at the temperature of Sixty-two Degrees by Fahrenheit's Thermometer, shall be and is hereby denominated the 'Imperial Standard Yard.'"

To provide for accidents to the bar, the Act continues: "And whereas it is expedient that the said Standard Yard, if lost, destroyed, defaced, or otherwise injured, should be restored to the same Length by reference to some invariable natural Standard: And whereas it has been ascertained by the Commissioners appointed by His Majesty to inquire into the subject of Weights and Measures, that the Yard hereby declared to be the Imperial Standard Yard, when compared with a Pendulum vibrating Seconds of Mean Time in the Latitude of London in a Vacuum at the Level of the Sea, is in the proportion of Thirty-six Inches to Thirty-nine Inches and one thousand three hundred and ninety-three ten-thousandth Parts of an Inch."

The new bar was made, however, not by this method, but by comparing several copies of the original and striking their average length. Four accurate duplicates of the new standard were secured, one of which is kept in the Mint, one in the charge of the Royal Society, one at Westminster Palace, and the fourth at the Royal Observatory, Greenwich. In addition, forty copies were distributed among the various foreign governments, all of the same metal as the original.

The French metre has also been standardised, being equal to one ten-millionth part of a quadrant of the earth's meridian (*i.e.* of the distance from the Equator to either of the Poles), that is, to 39·370788 inches. Professor A. A. Michelson has shown that any standard of length may be restored by reference to the measurement of wave lengths of light, with an error not exceeding one ten-millionth part of the whole.

It might be asked "Why should standards of such great accuracy be required?" In rough work, such as carpentry, it does not, indeed, matter if measurements are the hundredth of an inch or so out. But when we have to deal with scientific instruments, telescopes, measuring machines, engines for dividing distances on a scale, or even with metal turning, the utmost accuracy becomes needful; and a number of instruments will be much more alike in all dimensions if compared individually with a common standard than if they were only compared with one another. Supposing, for instance, a bar of exact diameter is copied; the copy itself copied; and so on a dozen times; the last will probably vary considerably from the correct measurements.

Hence it became necessary to standardise the foot and the inch by accurate subdivisions of the yard. This was accomplished by Sir Joseph Whitworth, who in 1834 obtained two standard yards in the form of measure bars, and by the aid of microscopes transferred the distance between the engraved lines to a rectangular *end*-measure bar, *i.e.* one of which the end faces are exactly a yard apart.

He next constructed his famous machine which is capable of detecting length differences of *one millionth* of an inch. Two bars are advanced towards each other by screw gearing: one by a screw having twenty threads to the inch, and carrying a graduated hand-wheel with 250 divisions on its rim; the other by a similar screw, itself driven by a worm-screw, working on the rim, which carries 200 teeth. The worm-screw has a hand-wheel with a micrometer graduation into 250 divisions of its circumference. So that, if this be turned one division, the second screw is turned only $\frac{1}{250} \times \frac{1}{200}$ of a division, and the bar it drives advances only $\frac{1}{20} \times \frac{1}{200} \times \frac{1}{250} = \frac{1}{1,000,000}$ of an inch. The screw at the other end of the machine (which in appearance somewhat resembles a metal lathe) is used for rapid adjustment only.

DELICATE MEASURING MACHINES

The upper illustration shows a Pratt-Whitney Measuring Machine in operation to decide the thickness of a cigarette paper, which is one-thousandth of an inch thick. This machine will measure variations of length or thickness as minute as one hundredth-thousandth of an inch. The lower illustration shows a Whitworth Measuring Machine which is sensitive to variations of one-millionth of an inch.

"He (Sir J. Whitworth) obtained the subdivision of the yard by making three foot pieces as nearly alike as was possible, and working these foot pieces down until each was equal to the others, and placing them end to end in his millionth measuring machine; the total length of the three foot pieces was then compared with a standard end-measure yard. These three foot pieces were ground until they were exactly equal to each other, and the three added together are equal to the standard yard. The subdivision of the foot into inch pieces was made in the same way."[2]

A doubt may have arisen in the reader's mind as to the possibility of determining whether the measuring machine is screwed up to the exact *tightness*. Would the measuring bars not compress a body a little before it appeared tight? Workmen, when measuring a bar with callipers, often judge by the sense of touch whether the jaws of the callipers pass the bar with the proper amount of resistance; but when one has to deal with millionths of an inch, such a method would not suffice. So Sir Joseph Whitworth introduced a *feeling-piece*, or *gravity-piece*. Mr. T. M. Goodeve thus describes it in *The Elements of Mechanism*: The gravity-piece consists of a small plate of steel with parallel plane sides, and having slender arms, one for its partial support, and the other for resting on the finger of the observer. One arm of the piece rests on a part of the bed of the machine, and the other arm is tilted up by the forefinger of the operator. The plane surfaces are then brought together, one on each side of the feeling-piece, until the pressure of contact is sufficient to hold it supported just as it remained when one end rested on the finger. This degree of tightness is perfectly definite, and depends on the weight of the gravity-piece, but not on the estimation of the observer.

In this way the expansion due to heat when a 36-inch bar has been touched for an instant with the finger-nail may be detected.

One of the most beautiful measuring machines commercially used comes from the factories of the Pratt-Whitney Co., Hartford, Connecticut, the well-known makers of machine tools and gauges of all kinds. It is made in different sizes, the largest admitting an 80-inch bar. Variations of $\frac{1}{100,000}$ of an inch are readily determined by the use of this machine. It therefore serves for originating gauge sizes, or for duplicating existing standards. The adjusting screw has fifty threads to the inch, and its index-wheel is graduated to 400 divisions, giving an advance of $\frac{1}{20,000}$ inch for each division: while by estimation this may be further subdivided to indicate one-half or even one-quarter of this small amount. Delicacy of contact between the measuring faces is obtained by the use of auxiliary jaws holding a small cylindrical gauge by the pressure of a light helical spring which operates the sliding spindle to which one of these auxiliary jaws is attached.

On one side of the "head" of the machine is a vertical microscope directed downwards on to a bar on the bed-plate, in which are a number of polished steel plugs graved with very fine central cross lines, each exactly an inch distant from either of its neighbours. A cross wire in the microscope tells when it is accurately abreast of the line below it. Supposing, then, that a standard bar three inches in diameter has to be tested. The "head" is slid along until the microscope is exactly over the "zero" plug line, and the divided index-wheel is turned until the two jaws press each other with the minimum force that will hold up the feeling-piece. Then the head is moved back and centred on the 3-inch line, and the bar to be tested is passed between the jaws. If the feeling-piece drops out it is too large, and the wheel is turned back until the jaws have been opened enough to let the bar through without making the feeling-piece fall. An examination of the index-wheel shows in hundred-thousandths of an inch what the excess diameter is.

On the other hand, if the bar were too small, the jaws would need to be closed a trifle: this amount being similarly reckoned.

We have now got into a region of very "practical politics," namely, the subject of *gauges*. All large engineering works which turn out machinery with interchangeable parts, *e.g.* screws and nuts, must keep their dimensions very constant if purchasers are not to be disgusted and disappointed. The small motor machinery so much in evidence to-day demands that errors should be kept within the ten-thousandth of an inch. An engineer therefore possesses a set of standard gauges to test the diameter and pitch of his screw threads and nuts; the size of tubes, wires; the circumference of wheels, etc.

Great inconvenience having been experienced by American railroad-car builders on account of the varying sizes of the screws and bolts which were used on the different tracks—though all were supposed to be of standard dimensions—the masters determined to put things right; and accordingly Professors Roger and Bond and the Pratt-Whitney Co. were engaged to work in collaboration in connection with the manufacture of tools for minute measurements, viz. to $\frac{1}{50,000}$ inch. "To give an idea of what is implied by this, let it be supposed that a person should take a pair of dividing compasses and lay off 50,000 prick-marks $\frac{1}{8}$ inch apart in a straight line. To do this the line would require to be over 520 feet, or nearly a tenth of a mile long. Imagine that many prick-marks compressed into the space of an inch, and you have an imperfect idea of the minuteness of the measurements which can now be made by the Pratt and Whitney Co."[3]

The standard taps and dies were supplied to tool-makers and engineers, who could thus determine whether articles supplied to them were of the proper dimensions. Nothing more was then heard of nuts being a "trifle small" or bolts "a leetle large." And so beautifully tempered were the dies made from the standards that one manufacturer claimed to have cut 18,800 cold-pressed nuts without any difference being perceptible in their sizes.

To appreciate what the difference of a thousandth of an inch makes in a true fit, you should handle a set of plug and ring gauges; the ring a true half-inch internally, the plugs half-inch, half an inch less one ten-thousandth of an inch, and half an inch less one-thousandth, in diameter.

The true half-inch plug needs to be forcibly driven into the ring on account of the friction between the surfaces. The next, if oiled, will slide in quite easily, but if left stationary a moment will "seize," and have to be driven out. The third will wobble very perceptibly, and would be at once discarded by a good workman as a bad fit.

For extremely accurate measurements of rods, calliper gauges, shaped somewhat like the letter Y, are used, the horns terminating in polished parallel jaws. Such a gauge will detect a difference of $\frac{1}{20,000}$ inch quite easily.

So accurately can plug gauges be made by reference to a measuring machine, that a gold leaf $\frac{1}{30,000}$ inch thick would be three times too thick to insert between the gauge and the jaws of the machine!

You must remember that in high-class workmanship these gauges are constantly being used. As time goes on, the "limit of error" allowed in many classes of machine parts is gradually lessened, which shows the simultaneous improvement of all machinery used in the handling of metal. James Watt was terribly hampered, when developing his steam-engine, by the difficulty of procuring a true cylinder for his pistons to work in with any approach to steam-tightness. His first cylinder was made by a smith of hammered iron soldered together. The next was cast and bored, but stuffing it with paper, cork, putty, pasteboard, and "old hat" proved useless to stem the leakage of steam. No wonder, considering that the finished cylinder was one-eighth of an inch larger in diameter at one end than at the other. Watt was in advance of his time. Neither machinery nor workmanship had progressed sufficiently to meet the requirements of the steam-engine. To-day an engineer would confidently undertake to bore a cylinder five feet in diameter with a variation from truth of not more than one five-hundredth of an inch.

Before passing from the subject of measuring machines, which play so important a part in modern mechanism, we may just glance at the electrical method of Dr. P. E. Shaw. He discovered recently that two clean metal surfaces can, by means of an electric current, feel one another on touching with a delicacy that far transcends that of the purely mechanical machine. The mechanism he employs is thus devised: A finely cut vertical screw having fifty threads to the inch has a disc graduated into 500 parts. The screw can be turned by means of a pulley string from a distance, and it is thus possible to give the top end of the screw a movement of $\frac{1}{25,000}$ inch, when a movement corresponding to one graduation is made.

This small movement is reduced by a train of six levers, the long arm of each bearing on the short arm of the one before it. The movement of the last lever of the train is thus reduced to $\frac{1}{4,000}$ of that of the screw point, so a movement of $\frac{1}{4,000} \times \frac{1}{25,000} \times = \frac{1}{1,00,000,000}$ inch is obtained!

How can such a movement be judged? A telephone and voltaic cell are joined to the last lever of the train and to the object whose movement is under examination. If they touch, the telephone sounds. An observer listens in the telephone, and if the object moves for any reason he can find out how much it moves by turning the screw until contact is made again.

Out of the many applications of this apparatus three may be given.

(1) A short bar of iron when magnetised elongates about $\frac{1}{1,000,000}$ of its length. If further magnetised it contracts. These changes can readily be measured with the instrument.

(2) The smallest sound audible in the telephone is due to a movement of the diaphragm of the telephone by about $\frac{1}{50,000,000}$ of an inch. This has been actually measured by Dr. Shaw and is by far the smallest distance ever directly recorded. It is about twice the diameter of the molecules of matter.

(3) Dispensing with levers, the screw alone is used for rougher work. Dr. Shaw has shown that one hundred-thousandth of an inch is the smallest dimension visible under a microscope. By fitting an electric measuring apparatus to the microscope carriage it becomes quite easy to measure minute distances. The microscope contains a cross wire which, when the object has been laid on the microscope stage, is centred on one side of the object. The

electric contact screw is then advanced till it makes contact with the stage and a sound arises in the telephone. A reading of the screw disc having been taken, the screw is drawn in and the microscope stage is traversed sufficiently to bring the wire in line with the other side of the object. Once more the operator makes electrical contact and gets a second reading, the difference between the two being the diameter of the object. In this manner the bacillus of tuberculosis has been proved to have an average diameter of $\frac{31}{250,000}$ of an inch.

The same method is employed to gauge the distance between the lines on a diffraction grating.

FOOTNOTES:

1. April 19th, 1901.

2. G. M. Bond in a lecture delivered before the Franklin Institute, February 29th, 1884.

3. *Report on Standard Screw Threads*, Philadelphia, 1884.

CHAPTER II

CALCULATING MACHINES

The simplest form of calculating machine was the Abacus, on which the schoolboys of ancient Greece did their sums. It consisted of a smooth board with a narrow rim, on which were arranged rows of pebbles, bits of bone or ivory, or silver coins. By replacing these little counters by sand, strewn evenly all over its surface, the abacus was transformed into a slate for writing or geometrical lessons. The Romans took the abacus, along with many other spoils of conquest, from the Greeks and improved it, dividing it by means of cross-lines, and assigning a multiple value to each line with regard to its neighbours. From their method of using the calculi, or pebbles, we derive our English verb, to *calculate*.

During the Middle Ages the abacus still flourished, and it has left a further mark on our language by giving its name to the Court of Exchequer, in which was a table divided into chequered squares like this simple school appliance.

Step by step further improvements were made, most important among them being those of Napier of Merchiston, whose logarithms vex the heads of our youth, and save many an hour's calculation to people who understand how to handle them. Sir Samuel Morland, Gunter, and Lamb invented other contrivances suitable for trigonometrical problems. Gersten and Pascal harnessed trains of wheels to their "ready-reckoners," somewhat similar to the well-known cyclometer.

All these devices faded into insignificance when Mr. Charles Babbage came on the scene with his famous calculator, which is probably the most ingenious piece of mechanism ever devised by the human brain. To describe the "Difference Engine," as it is called, would be impossible, so complicated is its character. Dr. Lardner, who had a wonderful command of language, and could explain details in a manner so lucid that his words could almost always be understood in the absence of diagrams, occupied twenty-five pages of the *Edinburgh Review* in the endeavour to describe its working, but gave several features up as a bad job. Another clever writer, Dr. Samuel Smiles, frankly shuns the task, and satisfies himself with the following brief description:—

"Some parts of the apparatus and modes of action are indeed extraordinary—and, perhaps, none more so than that for ensuring accuracy in the calculated results—the machine actually correcting itself, and rubbing itself back into accuracy, by the friction of the adjacent machinery! When an error is made the wheels become locked and refuse to proceed; thus the machine must go rightly or not at all—an arrangement as nearly resembling volition as anything that brass and steel are likely to accomplish." [4]

Mr. Babbage, in 1822, entered upon the task of superintending the construction of a machine for calculating and printing mathematical and astronomical tables. He began by building a model, which produced forty-four figures per minute. The next year the Royal Society reported upon the invention, which appeared so promising that the Lords of the Treasury voted Mr. Babbage £1,500 to help him perfect his apparatus.

He looked about for a first-rate mechanician of high intelligence as well as of extreme manual skill. The man he wanted appeared in Mr. Joseph Clement, who had already made his name as the inventor of a drawing instrument, a self-acting lathe, a self-centring chuck, and fluted taps and dies. Mr. Clement soon produced special tools for shaping the various parts of the machine. So elaborate was the latter, that, according to Dr. Smiles, "the drawings for the calculating machinery alone—not to mention the printing machinery, which was almost equally elaborate—covered not less than four hundred square feet of surface!"

You will easily imagine, especially if you have ever had a special piece of apparatus made for you by a mechanic, that the bills mounted up at an alarming rate; so fast, indeed, that the Government began to ask, Why this great expense, and so little visible result? After seven years' work the engineers' account had reached £7,200, and Mr. Babbage had disbursed an additional £7,000 out of his own pocket. Mr. Clement quarrelled with his employer—possibly because he harboured suspicions that they were both off on a wild-goose chase—and withdrew, taking all his valuable tools with him. The Government soon followed his example, and poor Babbage was left with his half-finished invention, "a beautiful fragment of a great work." It had been designed to calculate as far as twenty figures, but was completed only sufficiently to go to five figures. In 1862 it occupied a prominent place among the mechanical exhibits at the Great Exhibition.

A MECHANICAL CASHIER

The printing apparatus of a National Cash Register. It impresses on a paper strip the amount and nature of every money transaction; and also prints a date, number, advertisement, money value, and nature of business done on a ticket for the customer.

We learn, with some satisfaction, that all this effort was not fated to be fruitless. Two scientists of Stockholm—Scheutz by name—were so impressed by Dr. Lardner's account of this calculating machine that they carried Babbage's scheme through, and after twenty years of hard work completed a machine which seemed to be almost capable of thinking. The English Government spent £1,200 on a copy, which at Somerset House entered upon the routine duty of working out annuity and other tables for the Registrar-General.

From Babbage's wonderfully and fearfully made machine we pass to a calculator which to-day may be seen at work in hundreds of thousands of shops and offices.

It is the most modern substitute for the open till; and, by the aid of marvellous interior works, acts as account-keeper and general detective to the money transactions of the establishment in which it is employed.

There are very many types of Cash Register, and as it would be impossible to enumerate them all, we will pass at once to the most perfect type of all, known to the makers and vendors as "Number 95."

This register has at the top an oblong window. Dotted about the surface confronting the operator are, in the particular machine under notice, fifty-seven keys; six bearing the letters A, B, D, E, H, K; three the words "Paid out," "Charge," "Received on Account"; and the others money values ranging from £9 to $\frac{1}{4}$ d.

These are arranged in vertical rows. At the left end of the instrument is a printing apparatus, kept locked by the proprietor; at the right end a handle and a small lever. Below the register are six drawers, each labelled with an initial.

A customer enters the shop, and buys goods to the value of 6s. 11d. An assistant, to whom belongs the letter H, receives a sovereign in payment. He goes to the register, and after making sure that his drawer is pushed in till it is locked, first presses down the key H, and then the keys labelled "6s." and "11d."

Suddenly, like two Jacks-in-the-box, up fly into the window two tablets, with "6s. 11d." on both their faces, so that customer and assistant can see the figures. Simultaneously a bell of a certain tone rings, drawer H flies open (so that he may place the money in it and give change, if necessary), and a rotating arm in the window shows the word "cash."

The assistant now revolves the handle and presses the little lever. From a slot on the left side out flies a ticket, on the front of which is printed the date, a consecutive number, the assistant's letter, and the amount of the sale. The back has also been covered with an advertisement of some kind. The ticket and change are handed over to the customer, the drawer is shut, and the transaction is at an end, except for an entry in the shop's books of the article sold.

A carrier next comes in with a parcel on which five-pence must be paid for transport. Mr. A. receives the goods, goes to the register, presses his letter, the key with the words "paid out" on it, and the key carrying "5d.," takes out the amount wanted, and gives it to the carrier.

Again, a gentleman enters, and asks for change for half a sovereign. Mr. B. obliges him, pressing down his letter, but no figures.

Fourthly, a debtor to the shop pays five shillings to meet an account that has been against him for some time. Mr. K. receives the money and plays with the keys K, "Received on account," and "5s.," giving a ticket receipt.

Lastly, a customer buys a pair of boots on credit. Mr. D. attends to him, and though no cash is handled, uses the register, pressing the letter "Charge," and, say, "16s. 6d."

Now what has been going on inside the machine all this time? Let us lift up the cover, take off the case of the printing apparatus, and see.

A strip of paper fed through the printing mechanism has on it five rows of figures, letters, etc., thus—

		s.	d.
	H	6	11
Pd.	A	0	5
	B	0	0
Rc.	K	5	0
Ch.	D	16	6

The proprietor is, therefore, enabled to see at a glance (1) who served or attended to a customer, (2) what kind of business he did with him, (3) the monetary value of the transaction. At the end of the day each assistant sends in his separate account, which should tally exactly with the record of the machine.

Simultaneously with the strip printing, special counting apparatus has been (*a*) adding up the total of all money taken for goods, (*b*) recording the number of times the drawer has been opened for each purpose. Here, again, is a check upon the records.

This ingenious machine not only protects the proprietor against carelessness or dishonesty on the part of his employés, but also protects the latter against one another. If only one drawer and letter were used in common, it would be impossible to trace an error to the guilty party. The lettering system also serves to show which assistant does the most business.

Where a cash register of this type is employed every transaction must pass through its hands—or rather mechanism. It would be risky for an assistant not to use the machine, as eyes may be watching him. He cannot open his drawers without making a record; nor can he make a record without first closing the drawers; so that he must *give a reason* for each use of the register. If he used somebody else's letter, the ear of the rightful owner would at once be attracted by the note of his particular gong. When going away for lunch, or on business, a letter can be locked by means of a special key, which fits none of the other five locks.

The printing mechanism is particularly ingenious. Every morning the date is set by means of index-screws: and a consecutive numbering train is put back to zero. A third division accommodates a circular "electro" block for printing the advertisements, and a fourth division the figure wheels.

The turn given to the handle passes a length of the ticket strip through, a slot—prints the date, the number of the ticket, an advertisement on the back, the assistant's letter, the nature of the business done, and feeds the paper on to the figures which give the finishing touch. A knife cuts off the ticket, and a special lever shoots it out of the slot.

The National Cash Register Company, for prudential reasons, do not wish the details of the internal machinery to be described; nor would it be an easy task even were the permission granted. So we must imagine the extreme intricacy of the levers and wheels which perform all the tasks enumerated, and turn aside to consider the origin and manufacture of the register, which are both of interest.

The origin of the cash register is rather nebulous, because twenty-five years ago several men were working on the same idea. It first appeared as a practical machine in the offices of John and James Ritty, who owned stores and coalmines at Dayton, Ohio. James Ritty helped and largely paid for the first experiments. He needed a mechanical cashier for his own business, and says that, while on an ocean steamer *en route* to London the revolving machinery gave him the suggestion worked out, on his return to Dayton, in the first dial-machine. This gave way to the key-machine with its display tablet, or indicator, held up by a supporting bar moved back by knuckles on the vertical tablet rod.

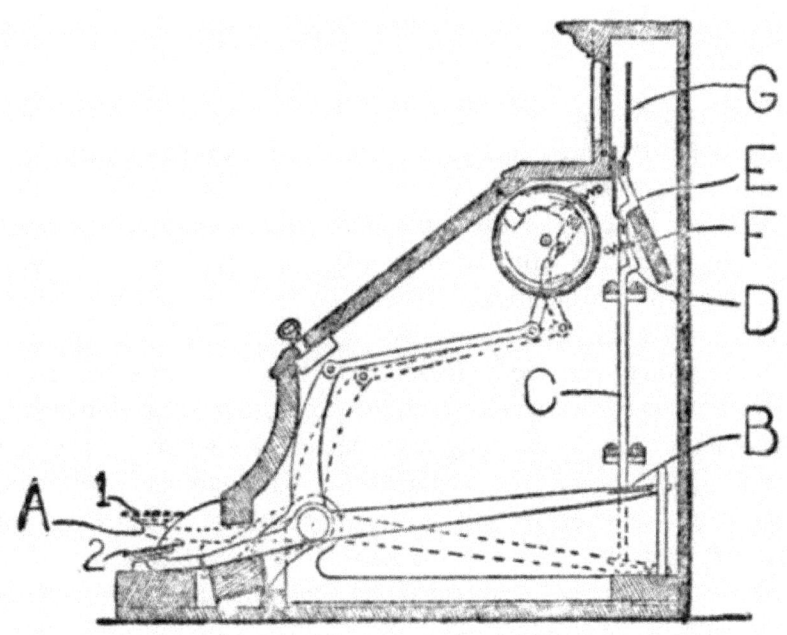

Fig. 1

The cut (Fig. 1) shows the right side of this key register, the action of which is thus described by the National Cash Register Company. The key A, when pressed with the finger at its ordinary position—marked 1—went down to the point marked 2. Being a lever and pivoted to its centre, pressing down a key elevated its extreme point B. This pushed up the tablet-rod C, having on its upper part the knuckle D. This knuckle D, pushed up, took the position at E; that is, the knuckle pushed back the supporting-bar F, and was pushed past it and held above it. If the same operation were performed on another key, the knuckle on its vertical rod, going up, would again push the supporting bar back, which would release the first knuckled rod, and leave the last one in its place. This knuckled rod had on its upper end the display tablet, or indicator G. James and John Ritty claimed and proved that they invented this, but the attorney for the Dayton Company (formed by them) in the Supreme Court was compelled to admit that this mechanism was old. Yet if machines built like this were exhibited elsewhere, they were at most only experimental models, and none of them had ever gone into practical or commercial use. In fact, at this time nothing had been really contributed which was useful to the public or used by the public.

The trouble was that the knuckles, being necessarily oiled, held dust and dirt which interfered with their free movement. And again, a "five-cent" or "ten-cent" key would be used more than others, and hence would become more worn. As a practical result the tablets did not drop when wanted, and the whole operation was thrown into confusion. When one tablet went up the other tablet stayed up, leaving a false indication. The most valuable modification now made by these Dayton inventors was to cease to rely on the knuckle to move back the supporting bar, and to supply the place of this function by what became known as "connecting mechanism," especially designed for this purpose. This was placed at the other, or say the left, side of the machine as you faced it. Cut No. 2 shows this new connecting mechanism. The keys, when pressed, performed the functions as before, on the right side of the machine, viz. to ring an alarm-bell, etc.; but on the other, or left, side the key, when pressed, operated the connecting mechanism marked M, N, O, P, and Q. The key pressed down by its leverage pushed back a little lever (Q), the further end of which pressed back the supporting bar F, and released the previously exposed indicator G, without relying on the knuckle to perform this function.

The Supreme Court of the United States said that the suggestion or idea to correct the old trouble and to drop the display tablet with certainty, and to accomplish this *by dividing the force used*, and applying a portion of it to the new connecting mechanism on the left side of the machine, "was fine invention," and that "the results are so important, and the ingenuity displayed to bring them about is such that we are not disposed to deny the patentees the merit of invention. The combination described in the first claim was clearly new."

To revert for a moment to the origin of the invention. Mr. John Ritty gives an account differing from that of his brother; but the two can probably be reconciled by supposing that the first ideas occurred simultaneously and were worked out in common.

Late one summer night, before dispersing home, a group of men were in his store. One of them said to the proprietor, "If you had a machine there to register the cash received, you would get more of it," and to the statement both owner and his clerks assented. This raised a laugh. But Ritty who, in spite of a large business, which ranged over everything from a needle to a haystack, did not make much profit by his sales, took the suggestion seriously, and put on his thinking-cap, with the result that the first machine was patented, and profits became very greatly increased.

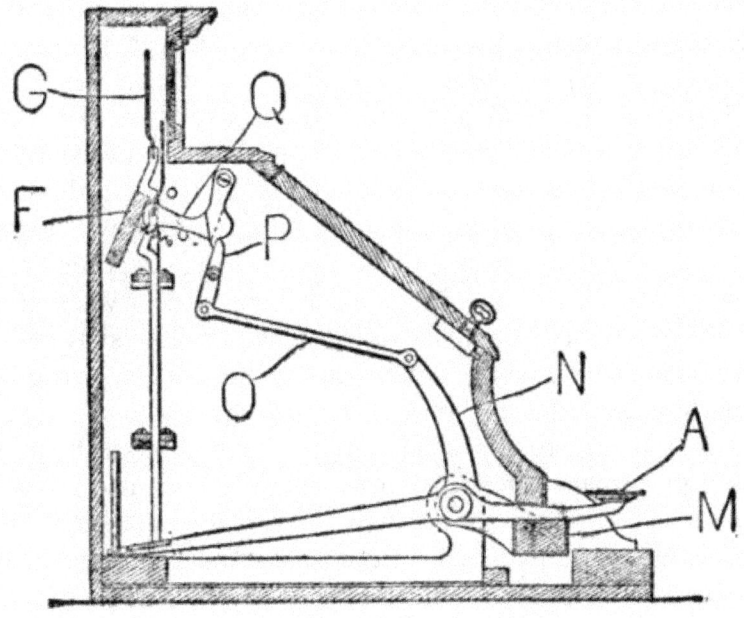

Fig. 2

Before his machine had been perfected a rival was in the field. Mr. Thomas Carney, a man who had seen much life as a lumber merchant, captain during the Civil War, explorer, and railroad promoter, settled down in 1884, at Chicago, to the manufacture of coin-changers. "When in various businesses," he says, "we used gold and silver only, and it seemed to be a sheer necessity to have something of a money-changer to assist us in handling it and making change. The custom then was to throw the different coins into a special receptacle marked for each. I invented, and in my own shop built this coin-changer, the keys of which, when touched, would, through the tube, drop the coin into the hand as wanted. At Chicago we made five or six hundred of these coin-changers, but by mistake placed the price too low, and after some conference I became assured that there was not enough money in it. A rich Chicago manufacturer had become familiar with the urgent need of a cash register, and the losses which followed in business without one. The National, at Dayton, had then been invented, but had not then been perfected as it has been since. Parties at Chicago agreed to put up the money if I would invent what would answer the purpose of a cash register and make a marketable machine. I went home and gave the matter some hard thinking, and talking with my son about the matter one night, I looked up at the clock and said, 'Why, Harry, there is the right thing. Sixty minutes make an hour; one hundred cents make a dollar. All I have got to do is to change the wheels a little, put some keys into it, and there will be a thing which will register cents, dimes, and dollars, just as that clock will register time in minutes and hours.' In clocks the minute wheel, when it has revolved to its sixty point, throws its added result of sixty minutes over on to another wheel, which takes up the story, with one hour in place of the old sixty minutes. The first wheel then begins again and goes its round. A second complete revolution of the minute wheel throws another sixty minutes on to the hour, and gives one more hour registered, making two hours, and so on. I took some wheels, and with pasteboard made hands and a machine. It was very rough, but I took it to my friends and explained it to them. We went on, but encountering difficulties and obstacles, we merged our whole enterprise in the National. I followed it, and have since invented, worked, and helped along in the National Cash Register service. I developed the No. 35 machine which the company began on and uses yet. It is now in use in every civilised country, for it can be made to register English money and any decimal currency."

In 1883 Dayton contained five families. The following year Colonel Robert Patterson bought a large property in the neighbourhood, and helped to develop a small town, which has since grown into a thriving manufacturing centre. His two sons, John H. Patterson and Frank J. Patterson, bought out all the original proprietors of the National Cash Register, greatly improved the machine's mechanism, and built the huge factory which employs about 4,000 men, women, and girls, and is one of the best-equipped establishments in the world to promote both an economical output and the comfort of the employés. The Company's buildings at Dayton cover 892,144 square feet of floor-space, and utilise 140 acres of ground. In convenience and attractiveness, and for light, heat, and ventilation, and all sanitary things, these structures are designed to be models of any used for factory purposes. A machine is made and sold every $2\frac{1}{2}$ minutes in the Dayton, Berlin, and Toronto factories collectively. According to its destination, it records dollars, shillings, marks, kronen, korona, francs, kroner, guildens, pesetas, pesos, milreis, rupees, or roubles. Registers are also made to meet the needs of the Celestials and the Japanese.

So necessary is it for these machines to be ever improving, that the Company, with a wisdom that prevails more largely, perhaps, in the United States than elsewhere, offer substantial rewards to the employé who records in a book kept specially for the purpose any suggestion which the committee, after due examination, consider likely to improve some detail of mechanism or manufacture. Five departments are entirely devoted to experiments carried out by a corps of inventors working with a special body of skilled mechanics. New patents accrue so fast as a result of this organised research that the National Company now owns 537 letters patent in the United States and 394 in foreign countries.

Many ideas come from outside. If they appear profitable they are bought and turned over to the Patents Department, which hands them on to the experimenters. These build an experimental model, which differs in many respects from the types hitherto manufactured. A cash register must be above all things strong, so that it can bear a heavy blow without getting out of order, and must retain its accuracy under all conditions.

The model finished, it goes before the inspectors, who thump it, hammer it, almost turn it inside out, and send it back to the Factory Committee with reports on any defects that may have come to light. If the inspectors can only knock the machine out of time they consider that they have done their duty; for they argue that, if weaknesses thus developed are put right, no purchaser will ever be able to dislocate the machinery if he stops short of an actual "brutal assault with violence."

Next comes the building of the commercial type, which will be sold by the thousand. The machine goes down to the tool-makers, a select board of seventy-five members, who list all the parts, and say how many drill-jigs, mills, fixtures, gauges, etc., are necessary to make every part. Then they draw out an approximate estimate of the cost of producing the tools, and after they have listed the parts, they turn them over to the various departments, such as the drafting-room, blacksmiths' shop, pattern shop, foundry, etc., after which the various parts are machined up. Then the tool-maker assembles together the various tools, and makes a number of the parts that each tool is designed for; so that when all the tools have done their preliminary work, the makers possess about fifty machines "in bits." These are assembled, to prove whether the tools do their business efficiently. If any part shows an inclination "to jam," or otherwise misbehave itself, the tool responsible is altered till its products are satisfactory.

Then, and only then—a period of perhaps two years may have elapsed since the model was first put in hand—the Company begins to entertain a prospect of getting back some of the money —any sum up to £50,000—spent in preparations. But they know that if people will only buy, they won't have much fault to find with their purchase. "Preparations brings success" is the motto of the N.C.R. So the Company spares no money, and is content to have £25,000 locked up in its automatic screw-making machines alone!

Human as well as inanimate machinery is well tended under the roof of the N.C.R. The committee believe that a healthy, comfortable employé means good—and therefore profitable—work; and that to work well, employés must eat and play well.

They therefore provide their boys with gardens, 10 feet wide by 170 feet in length; and pay an experienced gardener to direct their efforts. To encourage a start, bulbs, seeds, slips, etc., are supplied free; while prizes of considerable value help to stimulate competition.

One day, ten years or more ago, Mr. Patterson saw a factory girl trying to warm her tin bucket of cold coffee at the steam heater in the workshop. He is a humane man, and acting on the unintentional hint he built a lunch-room which contains, besides accommodation for 455 people, a piano and sewing-machine which the women can use during their noon recess of eighty minutes. A cooking school, dancing classes, and literary club are all available to members. The Company encourages its workers to own the houses they inhabit, and to make them as beautiful as their leisure will permit. Mr. Mosely, who took over to America an Industrial Commission of Experts in 1902, and an Educational Commission in the following year, paid visits on both occasions to the National Cash Register Works. In a speech to the Committee he said: "I do not know of any institution in the world which offers so beautiful an illustration of the proper working conditions as the National Cash Register Company. Your President has asked me to criticise. I cannot find anything to criticise in this factory. I have never seen such conditions in any other factory in the world, nor have I ever seen so many bright and intelligent faces as we have seen at luncheon in both the men's and women's dining rooms. I believe this factory is as nearly perfect as social conditions will permit."

NOTE.—The author desires to express his thanks to the National Cash Register Company for the kind help given him in the shape of materials for writing and illustrating this chapter.

FOOTNOTE:

4. *Industrial Biographies*, chap. xiii.

By permission of The Sphere.

The jacket of a 12-inch gun being turned in a monster lathe at Messrs. Vickers Maxim's works. Notice the long spiral strip coming off the edge of the cutting tool.

CHAPTER III

WORKSHOP MACHINERY

THE LATHE — PLANING MACHINES — THE STEAM HAMMER —
HYDRAULIC TOOLS — ELECTRICAL TOOLS IN THE SHIPYARD

"When I first entered this city," said Mr. William Fairbairn, in an inaugural address to the British Association at Manchester in 1861, "the whole of the machinery was executed by hand. There were neither planing, slotting, nor shaping machines, and with the exception of very imperfect lathes and a few drills, the preparatory operations of construction were effected entirely by the hands of the workmen. Now, everything is done by machine tools, with a degree of accuracy which the unaided hand could never accomplish. The automaton, or self-acting, machine tool has within itself an almost creative power; in fact, so great are its powers of adaptation, that there is no operation of the human hand that it does not imitate."

If such things could be said with justice forty-five years ago, what would Mr. Fairbairn think could he see the wonderful machinery with which the present-day workshop is equipped—machinery as relatively superior to the devices he speaks of as they were superior to the unaided efforts of the human hand? Invention never stands still. The wonder of one year is on the scrap-heap of abandoned machines almost before another twelve months have passed. Some important detail has been improved, to secure ease or economy in working, and a more efficient successor steps into its place. In his curious and original *Erewhon*, Mr. Samuel Butler depicts a community which, from the fear that machinery should become *too* ingenious, and eventually drain away man's capacity for muscular and mental action, has risen in revolt against the automaton, broken up all machines which had been in use for less than 270 years—with the exception of specimens reserved for the national museums—and reverted to hand labour. His treatment of the dangers attending the increased employment of lifeless mechanisms as a substitute for physical effort does not, however, show sympathy with the Erewhonians; since their abandonment of invention had obviously placed them at the mercy of any other race retaining the devices so laboriously perfected during the ages. And we, on our part, should be extremely sorry to part with the inanimate helpers which in every path of life render the act of living more comfortable and less toilsome.

So dependent are we on machinery, that we owe a double debt to the machines which create machines. A big factory houses the parents which send out their children to careers of usefulness throughout the world. We often forget, in our admiration of the offspring, the source from which they originated. Our bicycles, so admirably adapted to easy locomotion, owe their existence to a hundred delicate machines. The express engine, hurrying forward over the iron way, is but an assemblage of parts which have been beaten, cut, twisted, planed, and otherwise handled by mighty machines, each as wonderful as the locomotive itself. But then, we don't see these.

This and following chapters will therefore be devoted to a few peeps at the great tools employed in the world's workshops.

If you consider a moment, you will soon build up a formidable list of objects in which circularity is a necessary or desirable feature—wheels, shafts, plates, legs of tables, walking-sticks, pillars, parts of instruments, wire, and so on. The Hindu turner, whose assistant revolves with a string a wooden block centred between two short spiked posts let into the ground, while he himself applies the tool, is at one end of the scale of lathe users; at the other, we have the workman who tends the giant machine slowly shaping the exterior of a 12-inch gun, a propeller shaft, or a marble column. All aim at the same object—perfect rotundity of surface.

The artisans of the Middle Ages have left us, in beautiful balusters and cathedral screens, ample proofs that they were skilled workmen with the TURNING-LATHE. At the time of the Huguenot persecutions large numbers of French artificers crossed the Channel to England, bringing with them lathes which could cut intricate figures by means of wheels, eccentrics and other devices of a comparatively complicated kind. The French had undoubtedly got far ahead of the English in this branch of the mechanical arts, owing, no doubt, to the fact that the French *noblesse* had condescended to include turnery among their aristocratic hobbies.

With the larger employment of metal in all industries the need for handling it easily is increased. Much greater accuracy generally distinguishes metal as compared with woodwork. "In turning a piece of work on the old-fashioned lathe, the workman applied and guided his tool by means of muscular strength. The work was made to revolve, and the turner, holding the cutting tool firmly upon the long, straight, guiding edge of the rest, along which he carried it, and pressing its point firmly against the article to be turned, was thus enabled to reduce its surface to the required size and shape. Some dexterous turners were able, with practice and carefulness, to execute very clever pieces of work by this simple means. But when the article to be turned was of considerable size, and especially when it was of metal, the expenditure of muscular strength was so great that the workman soon became exhausted. The slightest variation in the pressure of the tool led to an irregularity of surface; and with the utmost care on the workman's part, he could not avoid occasionally cutting a little too deep, in consequence of which he must necessarily go over the surface again to reduce the whole to the level of that accidentally cut too deep, and thus possibly the job would be altogether spoiled by the diameter of the article under operation being made too small for its intended purpose."[5]

Any modern worker is spared this labour and worry by the device known as the SLIDE-REST. Its name implies that it at once affords a rigid support for the tool, and also the means of traversing the tool in a straight line parallel to the metal face on which work is being done.

The introduction of the slide-rest is due to the ingenuity of Mr. Henry Maudslay, who, at the commencement of the nineteenth century, was a foreman in the workshop of Mr. Joseph Bramah, inventor of the famous hydraulic press and locks which bear his name. His rest could be moved along the bed of the lathe by a screw, and clamped in any position desired. Fellow-workmen at first spoke derisively of "Maudslay's go-cart"; but men competent to judge its real value had more kindly words to say concerning it, when it had been adapted to machines of various types for planing as well as turning. Mr. James Nasmyth went so far as to state that "its influence in improving and extending the use of machinery has been as great as that produced by the improvement of the steam-engine in respect to perfecting manufactures and extending commerce, inasmuch as without the aid of the vast accession to our power of producing perfect mechanism which it at once supplied, we could never have worked out into practical and profitable forms the conceptions of those master minds who, during the last half century, have so successfully pioneered the way for mankind. The steam-engine itself, which supplies us with such unbounded power, owes its present perfection to this most admirable means of giving to metallic objects the most precise and perfect geometrical forms. How could we, for instance, have good steam-engines if we had not the means of boring out a true cylinder, or turning a true piston-rod, or planing a valve face? It is this alone which has furnished us with the means of carrying into practice the accumulated results of scientific investigation on mechanical subjects."

The screw-cutting lathe is so arranged that the slide-rest is moved along with its tool at a uniform speed by gear wheels actuated by the mechanism rotating the object to be turned. By changing the wheels the rate of "feed" may be varied, so that at every revolution the tool travels from $\frac{1}{64}$ of an inch upwards along the surface of its work. This regularity of action adds greatly to the value of the slide-rest; and the screw device also enables the workman to chase a thread of absolutely constant "pitch" on a metal bar; so that a screw-cutting lathe is not only a shaping machine but also the equivalent of a whole armoury of stocks and dies.

Some lathes have rests which carry several tools held at different distances from its axis, the cuts following one another deeper and deeper into the metal in a manner exactly similar to the harvesting of a field of corn by a succession of reaping machines. The recent improvements in tool-steel render it possible to get a much deeper cut than formerly, without fear of injury to the tool from overheating. This results in a huge saving of time.

For the boring of large cylinders an upright lathe is generally used, as the weight of the metal might cause a dangerous "sag" were the cylinder attached horizontally by one end to a facing-plate. Huge wheels can also be turned in this type of machine up to 20 feet or more in diameter; and where the cross-bar carrying the tools is fitted with several tool-boxes, two or more operations may be conducted simultaneously, such as the turning of the flange, the boring of the axle hole, and the facing of the rim sides.

A Gun Lathe. 154 feet long between centres, for boring and turning guns which, with their mountings, weigh 165 tons when complete. The makers are the Niles-Bement-Pond Co. of New York.

Perhaps the most imposing of all lathes are those which handle large cannon and propeller shafts, such as may be seen in the works of Sir W. G. Armstrong, Whitworth, and Company; of Messrs. Vickers, Sons and Maxim; and of other armament and shipbuilding firms. The Midvale Steel Company have in their shops at Hamilton, Ohio, a monster boring lathe which will take in a shaft 60 feet long, 30 inches in diameter, and bore a hole from one end to the other 14 inches in diameter. To do this, the lathe must attack the shaft at both ends simultaneously, as a single boring bar of 60 feet would not be stiff enough to keep the hole cylindrical. The shaft is placed in a revolving chuck in the central portion of the lathe—which has a total length of over 170 feet—and supported further by two revolving ring rests on each side towards the extremities. With work so heavy, the feeding up of the tool to its surface cannot be done conveniently by hand control, and the boring bars are therefore advanced by hydraulic pressure, a very ingenious arrangement ensuring that the pressure shall never become excessive.

Perhaps the type of lathe most interesting to the layman is the *turret* lathe, generally used for the manufacture of articles turned out in great numbers. The headstock—*i.e.* the revolving part which grips the object to be turned—is hollow, so that a rod may be passed right through it into the vicinity of the tools, which are held in a hexagon "turret," one tool projecting from each of its sides. When one tool has been finished with, the workman does not have the trouble of taking it out of the rest and putting another in its place; he merely turns the turret round, and brings another instrument opposite the work. If the object—say a water-cock—requires five operations performing on it in the lathe, the corresponding tools are arranged in their proper order round the turret. Stops are arranged so that as soon as any tool has advanced as far as is necessary a trip-action checks the motion of the turret, which is pulled back and given a turn to make it ready for the next attack.

One of the advantages of the turret lathe, particularly of the automatic form which shifts round the tool-box without human intervention, is its power of relieving the operator of the purely mechanical part of his work. Those who are familiar with the inside of some of our large workshops will have noticed men and boys who make the same thing all day and every day, and are themselves not far removed from machines. The articles they make are generally small and very rapidly produced, and the endless repetition of the same movements on the part of the operator is very tedious to watch, and must be infinitely more so to perform. Such an occupation is not elevating, and those engaged in it cannot take much interest in their work, or become fitted for a better position. When this work is done by an automatic lathe the machine performs the necessary operations, and the man supplies the intelligence, and, by exercising his thinking powers, becomes more valuable to his employers and himself. The introduction of new machines and methods generally has a stimulating effect on the whole shop, whatever the Erewhonians might say. The hubs and spindles of bicycles are cut from the solid bar by these automata; the tender has merely to feed them with metal, and they go on smoothing, shaping, and cutting off until the material is all used up. The existence of such lathes largely accounts for the low price of our useful metal steeds at the present time.

A great amount of shaping is now done by milling cutters in preference to firmly-fixed edged tools. The cutter is a rod or disc which has its sides, end, or circumference serrated with deep teeth, shaped to the section of the cut needed. Revolving at a tremendous speed, it quickly bites its way into anything it meets just so far as a stop allows it to go.

One of the most ingenious machines to which the milling tool has been fitted is the well-known Blanchard lathe, which copies, generally in wood, repetitive work, such as the stocks for guns and rifles. The lathe has two sets of centres—one for the copy, the other for the model—parallel on the same bed, and turned at equal speeds and in the same direction by a train of gear wheels. The milling cutter is attached to a frame, from which a disc projects, and is pressed by a spring against the model. As the latter revolves, its irregular shape causes the disc, frame, and cutter to move towards or away from its centre, and therefore towards or away from the centre of the copy, which has all superfluities whisked off by the cutter. The frame is gradually moved along the model, reproducing in the rough block a section similar to the part of the model which it has reached.

The self-centring chuck is an accessory which has proved invaluable for saving time. It may most easily be described as a circular plate which screws on to the inner end of the mandrel (the spindle imparting motion to the object being machined) and has in its face three slots radiating from the centre at angles of 120°. In each slot slides a stepped jaw, the under side of which is scored with concentric grooves engaging with a helical scroll turned by a key and worm gear acting on its circumference. The jaws approach or recede from the centre symmetrically, so that if a circular object is gripped, its centre will be in line with the axis of the lathe. Whether for gripping a tiny drill or a large wheel, the self-centring chuck is indispensable.

PLANING-MACHINES

Not less important in engineering than the truly curved surface is the true plane, in which, as Euclid would say, any two points being taken, the straight line between them lies wholly in that superficies. The lathe depends for its efficiency on the perfect flatness of all areas which should be flat—the guides, the surface plates, the bottom and sides of the headstock, and, above all, of the slide rest. For making plane metal superficies, a machine must first be constructed which itself is above suspicion; but when once built it creates machines like itself, capable of reproducing others *ad infinitum*.

Many amateur carpenters pride themselves on the beautiful smoothness of the boards over which they have run their jack planes. Yet, as compared with the bed of a lathe, their best work will appear very inaccurate.

The engineer's planing-machine in no way resembles its wooden relative. In the place of a blade projecting just a little way through a surface which prevents it from cutting too deep into the substance over which it is moving, we have a steel chisel very similar to the cutting tools of a lathe attached to a frame passing up and down over a bed to which the member holding the chisel is perfectly parallel. The article to be planed is rigidly attached to the bed and travels with it. Between every two strokes the tool is automatically moved sideways, so that no two cuts shall be in the same line. After the whole surface has been "roughed," a finishing cutter is brought in action, and the process is repeated with the business edge of the tool rather nearer to the bed.

Joseph Clement, a contemporary of Babbage, Maudslay, and Nasmyth, is usually regarded as the inventor of the planing-machine. By 1825 he had finished a planer, in which the tool was stationary and the work moving under it on a rolling bed. Two cutters were attached to the overhead cross rail, so that travel in either direction might be utilised. The bed of the machine, on which the work was laid, passed under the cutters on perfectly true rollers or wheels, lodged and held in their bearings as accurately as the best mandrel could be, and having set screws acting against their ends, totally preventing all end-motion. The machine was bedded on a massive and solid foundation of masonry in heavy blocks, the support at all points being so complete as effectually to destroy all tendency to vibration, with the object of securing full, round, and quiet cuts. The rollers on which the planing-machine travelled were so true, that Clement himself used to say of them, "If you were to put a paper shaving under one of the rollers it would at once stop the rest." Nor was this an exaggeration—the entire mechanism, notwithstanding its great size, being as true and accurate as a watch.[6] Mr. Clement next made a revolving attachment for the bed, in which bodies could be revolved under the cutter, on an axis parallel to the direction of travel. According to the wish of the operator, the object was converted into a cylinder, cone, or prism by its movements under the planing-tool. So efficient was the machine that it earned its maker upwards of ten pounds a day, at the rate of about eighteen shillings a square foot, until rivals appeared in the field and finally reduced the cost of planing to a few pence for the same area.

There are two main patterns of planes now in general use. The first follows the original design of Clement; the second has a fixed bed but a moving tool. Where the work is very heavy, as in the case of armour-plates for battleships, the power required to suddenly reverse the motion of a vast mass of metal is enormous, many times greater than the energy expended on the actual planing. For this reason the moving-bed machines have had to be greatly improved; and in some cases replaced by fixed-bed planers.

It is an impressive sight to watch one of these huge mechanisms reducing a rough plate, weighing twenty tons or more, to a smoothness which would shame the best billiard table. The machine, which towers thirty feet into the air and completely dwarfs the attendant, who has it as thoroughly under control as if it were a small file, bites great shining strips forty feet long, maybe, off the surface of the passive metal, and leaves a series of grooves as truly parallel as the art of man can make them. There is no fuss, no sticking, no stop, no noise; the force of electricity or steam, transmitted through wonderfully cut and arranged gear-wheels, is irresistible. The tool, so hard that a journey through many miles of steel has no appreciable effect on its edge, shears its way remorselessly over the surface which presently may be tempered to a toughness resembling its own. If you want to resharpen the tool, it will be no good to attack it with any known metal. But somewhere in the works there is a machine whose buzzing emery-wheels are more than a match for it, and rapidly grind the blunted edge into its former shape, so that it is ready to flay another plate, one skin at a time.

Planing-machines are of many shapes. Some have an upright on each side of the bed limiting the width of the work they can take; others are open-sided, one support of extra strength replacing the two, enabling the introduction of a plate twice as broad as the bed. Others, again, are built on the verge of a pit, so that they may cut the edges of an up-ended plate, and make it fit against its fellows so truly that you could not slip a sheet of paper edgeways between them. Thus has man, so frail and delicate in himself, shaped metal till it can torture its kind to suit his will, which he makes known to it by opening this valve or pulling on that lever. Not only does he flay it, but pierces it through and through; twists it into all manner of shapes; hacks masses off as easily as he would cut slices from a loaf; squeezes it in terrible presses to a fraction of its original thickness; and otherwise so treats it that we are glad that our scientific observations have as yet discovered no sentience in the substances reduced to our service.

THE STEAM HAMMER

The Scandinavian god Thor was a marvellous blacksmith. Thursday should remind us weekly of Odin's son, from whose hammer flashed the lightning; and, through him, of Vulcan, toiling at his smithy in the crater of Vesuvius. In spite of the pictures drawn for us by pagan mythologists of their god-smiths, we are left with the doubt whether these beings, if materialised, might not themselves be somewhat alarmed by the steam hammer which mere mortals wield so easily.

The forge is without dispute the "show-place" of a big factory, where huge blocks of metal feel the heavy hand of steam. As children we watched the blacksmith at his anvil, attracted and yet half-terrified by the spark-showers flying from a white-hot horseshoe. And even the adult, long used to startling sights, might well be fascinated and dismayed by the terrific blows dealt on glowing ingots by the mechanical sledge.

A steam hammer at work in Woolwich Arsenal, forging a steel ingot for the inner tube of a big gun. It delivers a blow equivalent to the momentum of a falling mass weighing 4000 tons. As speech is inaudible, the foreman gives hand signals to direct his men, who wear large canvas fingerless gloves to protect their hands from the intense heat.

James Nasmyth, the inventor of this useful machine, was the son of a landscape painter, who from his earliest youth had taken great interest in scientific and mechanical subjects of all kinds. At fifteen he made a steam-engine to grind his father's paints, and five years later a steam carriage "that ran many a mile with eight persons on it. After keeping it in action two months," he says in an account of his early life, "to the satisfaction of all who were interested in it, my friends allowed me to dispose of it, and I sold it—a great bargain—after which the engine was used in driving a small factory. I may mention that in that engine I employed the waste steam to cause an increased draught by its discharge up the chimney. This important use of waste steam had been introduced by George Stephenson some years before, though entirely unknown to me."

This interesting peep at the infancy of the motor carriage reveals mechanical capabilities of no mean order in young James. He soon entered the service of Mr. Joshua Field, Henry Maudslay's partner, and in 1834 set up a business on his own account at Manchester.

At this date the nearest approach to the modern steam hammer was the "tilt" hammer, operated by horse-, water-, or steam-power. It resembled an ordinary hand hammer on a very large scale, but as it could be raised only a small distance above its anvil, it became less effective as the size of the work increased, owing to the fall being "gagged." In 1837 Mr. Nasmyth interviewed the directors of the Great Western Steamship Company with regard to the manufacture of some unusually powerful tools which they needed for forging the paddle-shaft of the *Great Britain*. As the invention of the steam-engine had demanded the improvement of turning methods, so now the increase in the size of steamboats showed the insufficiency of forging machinery.

Mr. Nasmyth put on his thinking-cap. Evidently the thing

needed was a method for raising a very heavy mass of metal easily to a good height, so that its great weight might fall with crushing force on the object between it and the anvil. How to raise it? Brilliant idea! Steam! In a moment Nasmyth had mentally pictured an inverted steam cylinder rested on a solid upright overhanging the anvil and a block of iron attached to its piston-rod. All that would then be necessary was to admit steam to the under side of the piston until the block had risen to its full height, and to suddenly open a valve which would cut off the steam supply and allow the vapour already in the cylinder to escape.

By the next post he sent a sketch to the company, who approved his design heartily, but were unable to use it, since the need for the paddle-shaft had already been nullified by the substitution of a screw as the motive power of their ship. Poor Nasmyth knew that he had discovered a "good thing," but British forge-masters, with a want of originality that amounted to sheer blind stupidity, refused to look at the innovation. "We have not orders enough to keep in work the forge-hammers we have," they wrote, "and we don't want any new ones, however improved they may be."

His invention, therefore, appeared doomed to failure. Help, however, came from France in the person of Mr. Schneider, founder of the famous Creusot Iron Works, notorious afterwards as the birthplace of the Boer "Long Toms." Mr. Nasmyth happened to be away when Mr. Schneider and a friend called at the Manchester works, but his partner, Mr. Gaskell, showed the French visitors round the works, and also told them of the proposed steam hammer. The designs were brought out, so that its details might be clearly explained.

Years afterwards Nasmyth returned the visit, and saw in the Creusot Works a crank-shaft so large that he asked how it had been forged. "By means of your steam hammer," came the reply. You may imagine Nasmyth's surprise on finding the very machine at work in France which his own countrymen had so despised, and his delight over its obvious success.

On returning home he at once raised money enough to secure a patent, protected his invention, and began to manufacture what has been described as "one of the most perfect of artificial machines and noblest triumphs of mind over matter that modern English engineers have developed." A few weeks saw the first—a 30-cwt.—hammer at work. People flocked to watch its precision, its beauty of action, and the completeness of control which could arrest it at any point of its descent so instantaneously as to crack without smashing a nut laid on the anvil. "Its advantages were so obvious that its adoption soon became general, and in the course of a few years Nasmyth steam hammers were to be found in every well-appointed workshop both at home and abroad."[7]

Nasmyth's invention was improved upon in 1853 by Mr. Robert Wilson, his partner and successor. He added an automatic arrangement which raised the "tup," or head, automatically from the metal it struck, so that time was saved and loss of heat to the ingot was also avoided. The beauty of the "balance valve," as it was called, will be more clearly understood if we remember that the travel of the hammer is constantly increasing as the piece on the anvil becomes thinner under successive blows. Under the influence of this very ingenious valve every variety of blow could be dealt. By simply altering the position of a tappet lever by means of two screws, a blow of the exact force required could be repeated an indefinite number of times. "It became a favourite amusement to place a wine-glass containing an egg upon the anvil, and let the block descend upon it with its quick motion; and so nice was its adjustment, and so delicate its mechanism, that the great block, weighing perhaps several tons, could be heard playing tap, tap upon the egg without even cracking the shell, when, at a signal given to the man in charge, down would come the great mass, and the egg and glass would be apparently, as Walter Savage Landor has it, 'blasted into space.'"[8]

Later on Mr. Wilson added an equally important feature in the shape of a double-action hand-gear, which caused the steam to act on the top as well as the bottom of the piston, thus more than doubling the effect of the hammer.

The largest hammer ever made was that erected by the Bethlehem Iron Company of Pennsylvania. The "tup" weighed 125 tons. After being in use for three years the owners consigned it to the scrap-heap, as inferior to the hydraulic press for the manufacture of armour-plate, though it had cost them £50,000.

They then erected in its stead, for an equal sum of money, a 14,000-ton pressure hydraulic press, which fitly succeeds it as the most powerful of its kind in the world.

The change was made for three reasons. First, that the impact of so huge a block of metal necessitates the anvil being many times as heavy, and even then the shock to surrounding machinery may be very severe. Secondly, the larger the forging to be hammered, the less is the reaction of the anvil, so that all the force of the blow tends to be absorbed by the side facing the hammer; whereas with a small bar the anvil's inertia would have almost as much effect as the actual blow. Thirdly, the blow of the hammer is so instantaneous that the metal has not time to "flow" properly, and this leads to imperfect forgings, the surface of which may have been cracked. For very large work, therefore, the hammer is going out of fashion and the press coming in, though for lighter jobs it is still widely used.

Before leaving the subject we may glance at the double-headed horizontal hammer, such as is to be found in the forge-shop of the Horwich Railway Works. Two hammers, carried on rails and rollers, advance in unison from each side and pound work laid on a support between them. Each acts as anvil to the other, while doing its full share of the work. So that not only is a great deal of weight saved, but shocks are almost entirely absorbed; while the fact that each hammer need make a blow of only half the length of what would be required from a single hammer, enables twice as many blows to be delivered in a given time.

HYDRAULIC TOOLS

Before discussing these in detail we shall do well to trace the history of the Bramah press, which may be said to be their parent, since the principle employed in most hydraulic devices for the workshop, as also the idea of using water as a means of transmitting power under pressure, are justly attributed to Joseph Bramah.

If you take a dive into the sea and fall flat on the surface instead of entering at the graceful angle you intended, you will feel for some time afterwards as if an enemy had slapped you violently on the chest and stomach. You have learnt by sad experience that water, which seems to offer so little resistance to a body drawn slowly through it, is remarkably hard if struck violently. In fact, if enclosed, it becomes more incompressible than steel, without in any way losing its fluidity. We possess in water, therefore, a very useful agent for transmitting energy from one point to another. Shove one end of a column of water, and it gives a push to anything at its other end; but then it must be enclosed in a tube to guide its operation.

By a natural law all fluids press evenly on every unit of a surface that confines them. You may put sand into a bucket with a bottom of cardboard and beat hard upon the surface of the sand without knocking out the bottom. The friction between the sand particles and the bucket's sides entirely absorbs the blow. But if water were substituted for sand and struck with an object that just fitted the bucket so as to prevent the escape of liquid, the bottom, and sides, too, would be ripped open. The writer of this book once fired a candle out of a gun at a hermetically sealed tin of water to see what the effect would be. (Another candle had already been fired through an iron plate $\frac{1}{4}$ of an inch thick.) The impact *slightly* compressed the water in the tin, which gave back all the energy in a recoil which split the sheet metal open and flung portions of it many feet into the air. But the candle never got through the side.

This affords a very good idea of the almost absolute incompressibility of a liquid.

We may now return to history. Joseph Bramah was born in 1748 at Barnsley, in Yorkshire. As the son of a farm labourer his lot in life would probably have been to follow the plough had not an accident to his right ankle compelled him to earn his living in some other way. He therefore turned carpenter and developed such an aptitude for mechanics that we find him, when forty years old, manufacturing the locks with which his name is associated, and six years later experimenting with the hydraulic press. This may be described simply as a large cylinder in which works a solid piston of a diameter almost equal to that of the bore, connected to a force pump. Every stroke of the pump drives a little water into the cylinder, and as the water pressure is the same throughout, the total stress on the piston end is equal to that on the pump plunger multiplied by the number of times that the one exceeds the other in area. Suppose, then, that the plunger is one inch in diameter and the piston one foot, and that a man drives down the plunger with a force of 1,000 lbs., then the total pressure on the piston end will be 144 × 1,000 lbs.; but for every inch that the plunger has travelled the piston moves only $\frac{1}{144}$ of an inch, thus illustrating the law that what is gained in time is lost in power, and *vice versâ*.

The great difficulty encountered by Bramah was the prevention of leakage between the piston and the cylinder walls. If he packed it so tightly that no water could pass, then the piston jammed; if the packing was eased, then the leak recommenced. Bramah tried all manner of expedients without success. At last his foreman, Henry Maudslay—already mentioned in connection with the lathe slide-rest—conceived an idea which showed real genius by reason of its very simplicity. Why not, he said, let the water itself give sufficient tightness to the packing, which must be a collar of stout leather with an inverted U-shaped section? This suggestion saved the situation. A recess was turned in the neck of the cylinder at the point formerly occupied by the stuffing-box, and into this the collar was set, the edges pointing downwards. When water entered under pressure it forced the edges in different directions, one against the piston, the other against the wall of the recess, with a degree of tightness proportioned to the pressure. As soon as the pressure was removed the collar collapsed, and allowed the piston to pass back into the cylinder without friction. A similar device, to turn to smaller things for a moment, is employed in a cycle tyre inflater, a cup-shaped leather being attached to the rear end of the piston to seal it during the pressure stroke, though acting as an inlet valve for the suction stroke.

What we owe to Joseph Bramah and Henry Maudslay for their joint invention—the honour must be divided, like that of designing the steam hammer between Nasmyth and Wilson—it would indeed be hard to estimate. Wherever steady but enormous effort is required for lifting huge girders, houses, ships; for forcing wheels off their axles; for elevators; for advancing the boring shield of a tunnel; for compressing hay, wool, cotton, wood, even metal; for riveting, bending, drilling steel plates—there you will find some modification of the hydraulic press useful, if not indispensable.

However, as we are now prepared for a consideration of details, we may return to our workshop, and see what water is doing there. Outside stands a cylindrical object many feet broad and high, which can move up and down in vertical guides. If you peep underneath, you notice the shining steel shaft which supports the entire weight of this tank or coffer filled with heavy articles—stones, scrap iron, etc. The shaft is the piston-plunger of a very long cylinder connected by pipes to pumping engines and hydraulic machines. It and the mass it bears up serves as a reservoir of energy. If the pumping engines were coupled up directly to the hydraulic tools, whenever a workman desired to use a press, drill, or stamp, as the case might be, he would have to send a signal to the engine-man to start the pumps, and another signal to tell him when to stop. This would lead to great waste of time, and a danger of injuring the tackle from over driving. But with an accumulator there is always a supply of water under pressure at command, for as soon as the ram is nearly down, the engines are automatically started to pump it up again. In short, the accumulator is to hydraulic machinery what their bag is to bagpipes, or the air reservoir to an organ.

In large towns high-pressure water is distributed through special mains by companies who make a business of supplying factories, engineering works, and other places where there is need for it, though not sufficient need to justify the occupiers in laying down special pumping plant. London can boast five central distributing stations, where engines of 6,500 h.p. are engaged in keeping nine large accumulators full to feed 120 miles of pipes varying in diameter from seven inches downwards. The pressure is 700 lbs. to the square inch. Liverpool has twenty-three miles of pipes under 850 lbs. pressure; Manchester seventeen miles under 1,100 lbs. To these may be added Glasgow, Hull, Birmingham, Geneva, Paris, Berlin, Antwerp, and many other large cities in both Europe and the United States.

For very special purposes, such as making metal forgings,

pressures up to *twelve tons* to the square inch may be required. To produce this "intensifiers" are used, *i.e.* presses worked from the ordinary hydraulic mains which pump water into a cylinder of larger diameter connected with the forging press.

The largest English forging press is to be found in the Openshaw Works of Sir W. G. Armstrong, Whitworth, and Company. Its duty is to consolidate armour-plate ingots by squeezing, preparatory to their passing through the rolling mills. It has one huge ram 78 inches in diameter, into the cylinder of which water is pumped by engines of 4,000 h.p., under a pressure of 6,720 lbs. to the square inch, which gives a total ram force of 12,000 tons. It has a total height of 33 feet, is 22 feet wide, and 175 feet long, and weighs 1,280 tons. On each side of the anvil is a trench fitted with platforms and machinery for moving the ingot across the ingot block. Two 100-ton electric cranes with hydraulic lifting cylinders serve the press.

A HUGE HYDRAULIC PRESS

The 12,000-ton pressure Whitworth Hydraulic Press, used for consolidating steel ingots for armour-plating. Water is forced into the ram cylinder at a pressure of three tons to the square inch. Notice the man to the left of the press.

The Bethlehem Works "squeezer" has two rams, each of much smaller diameter than the Armstrong-Whitworth, but operated by a $10\frac{1}{2}$ tons pressure to the square inch. It handles ingots of over 120 tons weight for armour-plating. In 1895 Mr. William Corey, of Pittsburg, took out a patent for toughening nickel steel plates by subjecting them, while heated to a temperature of 2,000° F., to great compression, which elongates them only slightly, though reducing their thickness considerably. The heating of a large plate takes from ten to twenty hours; it is then ready to be placed between the jaws of the big press, which are about a foot wide. The plate is moved forward between the jaws after each stroke until the entire surface has been treated. At one stroke a 17-inch plate is reduced to 16 inches, and subsequent squeezings give it a final thickness of 14 inches. Its length has meanwhile increased from 16 to $18\frac{1}{2}$ feet, or in that proportion, while its breadth has remained practically unaltered. A simple sum shows that metal which originally occupied $32\frac{2}{3}$ cubic inches has now been compressed into 31 cubic inches. This alteration being effected without any injury to the surface, a plate very tough inside and very hard outside is made. The plate is next reheated to 1,350° F., and allowed to cool very gradually to a low temperature to "anneal" it. Then once again the furnaces are started to bring it back to 1,350°, when cold water is squirted all over the surface to give it a proper temper. If it bends and warps at all during this process, a slight reheating and a second treatment in the press restores its shape.

The hydraulic press is also used for bending or stamping plates in all manners of forms. You may see 8-inch steel slabs being quietly squeezed in a pair of huge dies till they have attained a semicircular shape, to fit them for the protection of a man-of-war's big-gun turret; or thinner stuff having its ends turned over to make a flange; or still slenderer metal stamped into the shape of a complete steel boat, as easily as the tinsmith stamps tartlet moulds. In another workshop a pair of massive jaws worked by water power are breaking up iron pigs into pieces suitable for the melting furnace.

The manufacture of munitions of war also calls for the aid of this powerful ally. Take the field-gun and its ammunition. "The gun itself is a steel barrel, hydraulically forged, and afterwards wire-wound; the carriage is built up of steel plates, flanged and shaped in hydraulic presses; the wheels have their naves composed of hydraulically flanged and corrugated steel discs, and even the tyres are forced on cold by hydraulic tyre-setters, the rams of which are powerful enough to reduce the diameter of the welded tyre until the latter tightly nips the wheel. The shells for the gun are punched and drawn by powerful hydraulic presses, and the copper driving-bands are fixed on the projectiles in special hydraulic presses. Quick-firing cartridge-cases are capped, drawn, and headed by an hydraulic press, whose huge mass always impresses the uninitiated as absurdly out of proportion to the small size of the finished case, and finally the cordite firing charge is dependent on hydraulic presses for its density and shape."[9]

The press for placing the "driving-band" on a shell is particularly interesting. After the shell has been shaped and its exterior turned smooth and true, a groove is cut round it near the rear end. Into this groove a band of copper is forced to prevent the leakage of gas from the firing charge past the shell, and also to bite the rifling which imparts a rotatory motion to the shell. The press for performing the operation has six cylinders and rams arranged spoke-wise inside a massive steel ring; the rams carrying concave heads which, when the full stroke is made, meet at the centre so as to form a complete circle. "Pressure is admitted," says Mr. Petch, "to the cylinders by copper pipes connected up to a circular distributing pipe. The press takes water from the 700-pounds main for the first $\frac{3}{8}$-inch of the stroke, and for the last $\frac{1}{8}$-inch water pressure at 3 tons per square inch is used. The total pressure on all the rams to band a 6-inch shell is only 600 tons, but for a 12-inch shell no less than 2,800 tons is necessary."

ELECTRIC TOOLS IN A SHIPYARD

Of late years electricity has taken a very prominent part in workshop equipment, on account of the ease with which it can be applied to a machine, the freedom from belting and overhead gear which it gives, and its greater economy. In a lathe-shop, where only half the lathes may be in motion at a time, the shafting and the belts for the total number is constantly whirling, absorbing uselessly a lot of power. If, however, a separate motor be fitted to each lathe, the workman can switch it on and off at his pleasure.

The New York Shipbuilding Company, a very modern

enterprise, depends mainly on electrical power for driving its machinery, in preference to belting, compressed air, or water. Let us stroll through the various shops, and note the uses to which the current has been harnessed. Before entering, our attention is arrested by a huge gantry crane, borne by two columns which travel on rails. From the cross girder, or bridge, 88 feet long, hang two lifting magnets, worked by 25 h.p. motors, which raise the load at the rate of 20 feet per minute. Motors of equal power move the whole gantry along its rails over the great piles of steel plates and girders from which it selects victims to feed the maw of the shops.

The main building is of enormous size, covering with its single roof no less than eighteen acres! Just imagine four acres of skylights and two acres of windows, and you may be able to calculate the little glazier's bill that might result from a bad hailstorm. In this immense chamber are included the machine, boiler, blacksmith, plate, frame, pipe, and mould shops, the general storerooms, the building ways, and outfitting slips. "The material which enters the plate and storage rooms at one end, does not leave the building until it goes out as a part of the completed ship for which it was intended, when the vessel is ready to enter service; there are installed in one main building, and under one roof, all the material and machinery necessary for the construction of the largest ship known to commerce, and eight sets of ship-ways, built upon masonry foundations, covered by roofs of steel and glass, and spanned by cranes up to 100 tons lifting capacity, are practically as much a part of the immense main building as the boiler shop or machine shop."[10]

A huge 100-ton crane of 121-foot span dominates the machine-shop and ship-ways at a height of 120 feet. It toys with a big engine or boiler, picking it up when the riveters, caulkers, and fitters have done their work, and dropping it gently into the bowels of a partly-finished vessel. A number of smaller cranes run about with their loads. Those which handle plates are, like the big gantry already referred to, equipped with powerful electro-magnets which fix like leeches on the metal, and will not let go their hold until the current is broken by the pressing of a button somewhere on the bridge. Sometimes several plates are picked up at once, and then it is pretty to see how the man in charge drops them in succession, one here, another there, by merely opening and closing the switch very quickly, so that the plate furthest from the magnets falls before the magnetism has passed out of the nearer plates.

Another interesting type is the extension-arm crane, which shoots out an arm between two pillars, grips something, and pulls it back into the main aisle, down which it travels without impediment.

On every side are fresh wonders. Here is an immense rolling machine, fed with plates 27 feet wide, which bends the $1\frac{1}{8}$-inch thick metal as if it were so much pastry; or turns over the edges neatly at the command of a 50 h.p. motor. There we have an electric plate-planer scraping the surface of a sheet half the length of a cricket pitch. As soon as a stroke is finished the bed reverses automatically, while the tool turns over to offer its edge to the metal approaching from the other side. All so quietly, yet irresistibly done!

Now mark these punches as they bite $1\frac{1}{4}$-inch holes through steel plates over an inch thick, one every two seconds. A man cutting wads out of cardboard could hardly perform his work so quickly and well. Almost as horribly resistless is the circular saw which eats its way quite unconcernedly through bars six inches square, or snips lengths off steel beams.

What is that strange-looking machine over there? It has three columns which move on circular rails round a table in the centre. Up and down each column passes a stage carrying with it a workman and an electric drill working four spindles. Look! here comes a crane with a boiler shell, the plates of which have been bolted in position. The crane lets down its load, end-up, on to the table, and trots off, while the three workmen move their columns round till the twelve drills are opposite their work. Then whirr! a dozen twisted steel points, ranged in three sets of four, one drill above the other, bite into the boiler plates, opening out holes at mathematically correct intervals all down the overlapping seam-plates. This job done, the columns move round the boiler, and their drills pierce it first near the lower edge, then near the upper. The crane returns, grips the cylinder, and bears it off to the riveters, who are waiting with their hydraulic presses to squeeze the rivets into the holes just made, and shape their heads into neat hemispheres. As it swings through the air the size of the boiler is dwarfed by its surroundings; but if you had put a rule to it on the table you would have found that it measured 20 feet in diameter and as many in length. A few months hence furnaces will rage in its stomach, and cause it to force tons of steam into the mighty cylinders driving some majestic vessel across the Atlantic.

We pass giant lathes busy on the propeller shafts, huge boring mills which slowly smooth the interior of a cylinder, planers which face the valve slides; and we arrive, eye-weary, at the launching-ways where an ocean liner is being given her finishing touches. Then we begin to moralise. That 600-foot floating palace is a concretion of parts, shaped, punched, cut, planed, bored, fixed by electricity. Where does man come in? Well, he harnessed the current, he guided it, he said "Do this," and it did it. Does not that seem to be his fair share of the work?

FOOTNOTES:

5. *Industrial Biographies*, Dr. S. Smiles.

6. *Industrial Biographies*.

7. *Industrial Biographies*.

8. *Chambers's Encyclopædia*.

9. Mr. A. F. Petch in *Cassier's Magazine*.

10. *Cassier's Magazine*.

CHAPTER IV

PORTABLE TOOLS

"IF the mountain won't come to Mahomet," says the proverb, "Mahomet must go to the mountain."

This is as true in the workshop as outside;—Mahomet being the tool, the mountain the work on which it must be used. With the increase in size of machinery and engineering material, methods half a century old do not, in many cases, suffice; especially at a time when commercial competition has greatly reduced the margin of profits formerly expected by the manufacturer.

To take the case of a large shaft, which must have a slot cut along it on one side to accommodate the key-wedge, which holds an eccentric for moving the steam valves of a cylinder, or a screw-propeller, so that it cannot slip. The mass weighs, perhaps, twenty tons. One way of doing the job is to transport the shaft under a drill that will cut a hole at each end of the slot area, and then to turn it over to the planer for the intermediate metal to be scraped out. This is a very toilsome and expensive business, entailing the use of costly machinery which might be doing more useful work, and the sacrifice of much valuable time. Inventors have therefore produced portable tools which can perform work on big bodies just as efficiently as if it had been done by larger machinery, in a fraction of the time and at a greatly reduced cost. To quote an example, the cutting of a key-way of the kind just described by big machines would consume perhaps a whole day, whereas the light, portable, easily attached miller, now generally used, bites it out in ninety minutes.

PNEUMATIC TOOLS

The best known of these is the pneumatic hammer. It consists of a cylinder, inside which moves a solid piston having a stroke of from half an inch to six inches. Air is supplied through flexible tubing from a compressing pump worked by steam. The piston beats on a loose block of metal carried in the end of the tool, which does the actual striking. The piston suddenly decreases in diameter at about the centre of its length, leaving a shoulder on which air can work to effect the withdrawal stroke. By a very simple arrangement of air-ports the piston is made to act as its own valve. As the plane side of the piston has a greater area than that into which the piston-rod fits, the striking movement is much more violent than the return. Under a pressure of several hundreds of pounds to the square inch a pneumatic hammer delivers upwards of 7,000 blows per minute; the quick succession of comparatively gentle taps having the effect of a much smaller number of heavier blows. For the flat hammer head can be substituted a curved die for riveting, or a chipping chisel, or a caulking iron, to close the seams of boilers.

The riveter is peculiarly useful for ship and bridge-building work where it is impossible to apply an hydraulic tool. A skilled workman will close the rivet heads as fast as his assistant can place them in their holes; certainly in less than half the time needed for swing-hammer closing.

Even more effective proportionately is the pneumatic chipper. The writer has seen one cut a strip off the edge of a half-inch steel plate at the rate of several inches a minute. To the uninitiated beholder it would seem impossible that a tool weighing less than two stone could thus force its way through solid metal. The speed of the piston is so high that, though it scales but a few pounds, its momentum is great enough to advance the chisel a fraction of an inch, and the individual advances, following one another with inconceivable rapidity, soon total up into a big cut.

Automatic chisels are very popular with ornamental masons, as they lend themselves to the sculpturing of elaborate designs in stone and marble.

Their principle, modified to suit work of another character, is seen in percussive rock drills, such as the Ingersoll Sergeant. In this case the piston and tool are solid, and the air is let into the cylinder by means of slide valves operated by tappets which the piston strikes during its movements. Some types of the rock-drill are controllable as to the length of their stroke, so that it can be shortened while the "entry" of the hole is being made and gradually increased as the hole deepens. For perpendicular boring the drill is mounted on a heavily weighted tripod, the inertia of which effectively damps all recoil from the shock of striking; for horizontal work, and sometimes for vertical, the support is a pillar wedged between the walls of the tunnel, or shaft. An ingenious detail is the rifled bar which causes the drill to rotate slightly on its axis between every two strokes, so that it may not jam. The drills are light enough to be easily erected and dismantled, and compact, so that they can be used in restricted and out-of-the way places, while their simplicity entails little special training on the part of the workman. With pneumatic and other power-drills the cost of piercing holes for explosive charges is reduced to less than one-quarter of that of "jumping" with a crowbar and sledgehammers. With the hand method two men are required, usually more; one man to hold, guide, and turn the drill; and the other, or others, to strike the blows with hammers. The machine, striking a blow far more rapidly than can be done by hand, reduces the number of operators to one man, and perhaps his helper. So durable is the metal of these wonderful little mechanisms that the delivery of 360,000 blows daily for months, even though each is given with a force of perhaps half a ton, fails to wear them out; or at the most only necessitates the renewal of some minor and cheap part. The debt that civilisation owes to the substitution of mechanical for hand labour will be fully understood by anyone who is conversant with the history of tunnel-driving and mining.

Another application of pneumatics is seen in the device for cutting off the ends of stay bolts of locomotive boilers. It consists of a cylinder about fifteen inches in diameter, the piston of which operates a pair of large nippers capable of shearing half-inch bars. The whole apparatus weighs but three-quarters of a hundredweight, yet its power is such that it can trim bolts forty times as fast as a man working with hammer and cold-chisel, and more thoroughly.

Then there is the machine for breaking the short bolts which hold together the outer and inner shells of the water-jacket round a locomotive furnace. A threaded bar, along which travels a nut, has a hook on its end to catch the bolt. The nut is screwed up to make the proper adjustment, and a pneumatic cylinder pulls on the hook with a force of many tons, easily shearing through the bolt.

We must not forget the *pneumatic borer* for cutting holes in wood or metal, or enlarging holes already existing. The head of the borer contains three little cylinders, set at an angle of 120°, to rotate the drill, the valves opening automatically to admit air at very high pressures behind the pistons. Any carpenter can imagine the advantage of a drill which has merely to be forced against its work, the movement of a small lever by the thumb doing the rest!

Next on the list comes the *pneumatic painter*, which acts on much the same principle as the scent-spray. Mechanical painting first came to the fore in 1893, when the huge Chicago Exposition provided many acres of surfaces which had to be protected from the weather or hidden from sight. The following description of one of the machines used to replace hand-work is given in *Cassier's Magazine*: "The paint is atomized and sprayed on to the work by a stream of compressed air. From a small air-compressor the air is led, through flexible hose, to a paint-tank, which is provided with an air-tight cover and clamping screws. The paint is contained in a pot which can be readily removed and replaced by another when a different colour is required. This arrangement of interchangeable tins is also important as facilitating easy cleaning. The container is furnished with a semi-rotary stirrer, the spindle passing through a stuffing-box in the cover, and ending in a handle by which the whole thing complete may be carried about. The compressor is necessarily fixed or stationary, but the paint-tank, connected to it by the single air-hose, can be moved close to the work, while the length of hose from the tank to the nozzle gives the freedom of movement necessary. Air-pressure is admitted to the tank by a bottom valve, and forces the paint up an internal pipe and along a hose from the tank to the spraying nozzle, to which air-pressure is also led by a second hose. The nozzle is practically an injector of special form. The flow of paint at the nozzle is controlled by a small plug valve and spring lever, on which the operator keeps his thumb while working, and which, on release, closes automatically. When it is required to change from one colour to another, or to use a different material, such as varnish, the can, previously in use, is removed, and air, or, if necessary, paraffin oil, is blown through the length of hose which supplies the paint until it is completely clean." The writer then mentions as an instance of the machine's efficiency that it has covered a 30 feet by 8 feet boiler in less than an hour, and that at one large bridge yard a 70 feet by 6 feet girder with all its projecting parts was coated with boiled oil in two hours—a job which would have occupied a man with a brush a whole day to execute. Apart from saving time, the machine produces a surface quite free from brush marks, and easily reaches surfaces in intricate mouldings which are difficult to get at with a brush.

The *pneumatic sand-jet* is used for a variety of purposes: for cleaning off old paint, or the weathered surface of stonework; for polishing up castings and forgings after they have been brazed. At the cycle factory you will find the sand-jet hard at work on the joints of cycle frames, which must be cleared of all roughness before they are fit for the enameller. The writer, a few days before penning these lines, watched a jet removing London grime from the face of a large hotel. Down a side street stood a steam-engine busily compressing air, which was led by long pipes to the jet, situated on some lofty scaffolding. The rapidity with which the flying grains scoured off smoke deposits attracted the notice of a large crowd, which gazed with upturned heads at the whitened stones. A peculiarity about the jet is that it proves much more effective on hard material than on soft, as the latter, by offering an elastic surface, robs the sand of its cutting power.

After merely mentioning the *pneumatic rammer* for forcing sand into foundry moulds, we pass to the *pneumatic sand-papering* machine, which may be described briefly as a revolving disc carrying a circle of sand-paper on its face revolved between guards which keep it flat to its work. The disc flies round many hundreds of times per minute, rapidly wearing down the fibrous surface of the wood it touches. When the coarse paper has done its work a finely-grained cloth is substituted to produce the finish needful for painting.

CHAPTER V

THE PEDRAIL: A WALKING STEAM-ENGINE

Have you ever watched carefully a steam-roller's action on the road when it is working on newly laid stones? If you have, you noticed that the stones, gravel, etc., in front of the roller moved with a wave-like motion, so that the engine was practically climbing a never-ending hill. No wonder then that the mechanism of such a machine needs to be very strong, and its power multiplied by means of suitable gearing.

Again, suppose that an iron-tyred vehicle, travelling at a rapid pace, meets a large stone, what happens? Either the stone is forced into the ground or the wheel must rise over it. In either case there will be a jar to the vehicle and a loss of propulsive power. Do not all cyclists know the fatigue of riding over a bumpy road—fatigue to both muscles and nerves?

As regards motors and cycles the vibration trouble has been largely reduced by the employment of pneumatic tyres, which *lap over* small objects, and when they strike large ones minimise the shock by their buffer-like nature. Yet there is still a great loss of power, and if pneumatic-tyred vehicles suffer, what must happen to the solid, snorting, inelastic traction-engine? On hard roads it rattles and bumps along, pulverising stones, crushing the surface. When soft ground is encountered, in sink the wheels, because their bearing surface must be increased until it is sufficient to carry the engine's weight. But by the time that they are six inches below the surface there will be a continuous vertical belt of earth six inches deep to be crushed down incessantly by their advance.

How much more favourably situated is the railway locomotive or truck. *Their* wheels touch metal at a point but a fraction of an inch in length; consequently there is nothing to hamper their progression. So great is the difference between the rail and the road that experiment has shown that, whereas a pull of from 8 to 10 lbs. will move a ton on rails, an equal weight requires a tractive force of 50 to 100 lbs. on the ordinary turnpike.

In order to obviate this great wastage of power, various

attempts have been made to provide a road locomotive with means for laying its own rail track as it proceeds. About forty years ago Mr. Boydell constructed a wheel which took its own rail with it, the rails being arranged about the wheel like a hexagon round a circle, so that as the wheel moved it always rested on one of the hexagon's sides, itself flat on the ground. This device had two serious drawbacks. In the first place, the plates made a rattling noise which has been compared to the reports of a Maxim gun; secondly, though the contrivance acted fairly well on level ground, it failed when uneven surfaces were encountered. Thus, if a brick lay across the path, one end of a plate rested on the brick, the other on the ground behind, and the unsupported centre had to carry a sudden, severe strain. Furthermore, the plates, being connected at the angles of the hexagon, could not tilt sideways, with the result that breakages were frequent.

Of late years another inventor, Mr. J. B. Diplock, has come forward with an invention which bids fair to revolutionise heavy road traffic. At present, though it has reached a practical stage and undergone many tests satisfactorily, it has not been made absolutely perfect, for the simple reason that no great invention jumps to finality all at once. Are not engineers still improving the locomotive?

The Pedrail, as it has been named, signifies a rail moving on feet. Mr. Diplock, observing that a horse has for its weight a tractive force much in excess of the traction-engine, took a hint from nature, and conceived the idea of copying the horse's foot action. The reader must not imagine that here is a return to the abortive and rather ludicrous attempts at a walking locomotive made many years ago, when some engineers considered it proper that a railway engine should be *propelled* by legs. Mr. Diplock's device not merely propels, but also steps, *i.e.* selects the spot on the ground which shall be the momentary point at which propulsive force shall be exerted. To make this clearer, consider the action of a wheel. First, we will suppose that the spokes, any number you please, are connected at their outer ends by flat plates. As each angle is passed the wheel falls flop on to the next plate. The greater the number of the spokes, the less will be each successive jar (or step); and consequently the perfect wheel is theoretically one in which the sides have been so much multiplied as to be infinitely short.

A horse has practically two wheels, its front legs one, its back legs the other. The shoulder and hip joints form the axles, and the legs the spokes. As the animal pulls, the leg on the ground advances at the shoulder past the vertical position, and the horse would fall forwards were it not for the other leg which has been advanced simultaneously. Each step corresponds to our many-sided wheel falling on to a flat side—and the "hammer, hammer, hammer on the hard high road" is the horsey counterpart of the metallic rattle.

On rough ground a horse has a great advantage over a wheeled tractor, because it can put its feet down *on the top* of objects of different elevations, and *still pull*. A wheel cannot do this, and, as we have seen, a loss of power results. Our inventor, therefore, created in his pedrail a compromise between the railway smoothness and ease of running and the selective and accommodating powers of a quadruped.

We must now plunge into the mechanical details of the pedrail, which is, strictly speaking, a term confined to the wheel alone. Our illustration will aid the reader to follow the working of the various parts.

In a railway we have (*a*) sleepers, on the ground, (*b*) rails attached to the sleepers, (*c*) wheels rolling over the rails. In the pedrail the order, reckoning upwards, is altered. On the ground is the *ped*, or movable sleeper, carrying wheels, over which a rail attached to the moving vehicle glides continuously. The *principle* is used by anyone who puts wooden rollers down to help him move heavy furniture about.

Of course, the peds cannot be put on the ground and left behind; they must accompany their rollers and rails. We will endeavour to explain in simple words how this is effected.

To the axles of the locomotive is attached firmly a flat, vertical plate, parallel to the sides of the fire-box. Pivoted to it, top and bottom, at their centres, are two horizontal rocking arms; and these have their extremities connected by two bow-shaped bars, or cams, their convex edges pointing outwards, away from the axle. Powerful springs also join the rocking arms, and tend to keep them in a horizontal position. Thus we have a powerful frame, which can oscillate up and down at either end. The bottom arm is the rail on which the whole weight of the axle rests.

The rotating and moving parts consist of a large, flat, circular case, the sides of which are a few inches apart. Its circumference is pierced by fourteen openings, provided with guides, to accommodate as many short sliding spokes, which are in no way attached to the main axle. Each spoke is shaped somewhat like a tuning-fork. In the **V** is a roller-wheel, and at the tip is a "ped," or foot. As the case revolves, the tuning-fork spokes pass, as it were, with a leg on each side of the framework referred to above; the wheel of each spoke being the only part which comes into contact with the frame. Strong springs hold the spokes and rollers normally at an equal distance from the wheel's centre.

It must now be stated that the object of the framework is to thrust the rollers outwards as they approach the ground, and slide them below the rail. The side-pieces of the frame are, as will be noticed (see Fig. 3), eccentric, *i.e.* points on their surfaces are at different distances from the axle centre. This is to meet the fact that the distance from the axle to the ground is greater in an oblique direction than it is vertically, and therefore for three spokes to be carrying the weight at once, two of them must be more extended than the third. So then a spoke is moved outward by the frame till its roller gets under the rail, and as it passes off it it gradually slides inwards again.

It will be obvious to the reader that, if the "peds" were attached inflexibly to the ends of their spokes they would strike the ground at an angle, and, of course, be badly strained. Now, Mr. Diplock meant his "peds" to be as like feet as possible, and come down *flat*. He therefore furnished them with ankles, that is, ball-and-socket joints, so that they could move loosely on their spokes in all directions; and as such a contrivance must be protected from dust and dirt, the inventor produced what has been called a "crustacean joint," on account of the resemblance it bears to the overlapping armour-plates of a lobster's tail. The plates, which suggest very thin quoits, are made of copper, and can be renewed at small cost when badly worn. An elastic spring collar at the top takes up all wear automatically, and renders the plates noiseless. This detail cost its inventor much work. The first joint made represented an expenditure of £6; but now, thanks to automatic machinery, any number can be turned out at 3s. 6d. each.

A word about the feet. A wheel has fourteen of these. They are eleven inches in diameter at the tread, and soled with rubber in eight segments, with strips of wood between the segments to prevent suction in clay soil. The segments are held together by a malleable cast-iron ring around the periphery of the feet and a tightening core in the centre. These wearing parts, being separate from the rest of the foot, are easily and cheaply renewed, and repairs can be quickly effected, if necessary, when on the road. The surface in contact with the ground being composed of the three substances—metal, wood, and rubber, which all take a bearing, provides a combination of materials adapted to the best adhesion and wear on any class of road, or even on no road at all.

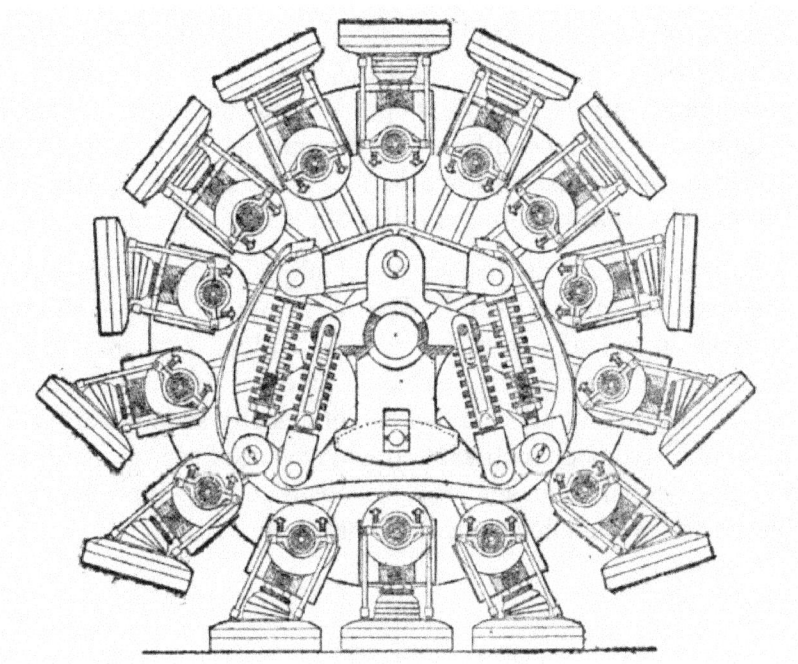

Fig. 3

Motive power is transmitted by the machinery to the wheel axle, from that to the casing, from the casing to the sliding spokes. As there are alternately two and three feet simultaneously in contact with the ground, the power of adhesion is very great—much greater than that of an ordinary traction-engine. This is what Professor Hele-Shaw says in a report on a pedrail tractor: "The weight of the engine is spread over no less than twelve feet, each one of which presses upon the ground with an area immensely greater—probably as much as ten times greater—than that of all the wheels (of an ordinary traction-engine) taken together on a hard road. Upon a soft road all comparison between wheels and the action of these feet ceases. The contact of each of the feet of the Pedrail is absolutely free from all slipping action, and attains the absolute ideal of working, being merely placed in position without sliding to take up the load, and then lifted up again without any sliding to be carried to a new position on the road."

It is necessary that the feet should come down flat on the ground. If they struck it at all edgeways they would "sprain their ankles"; otherwise, probably break off at the ball joint. Mechanism was, therefore, introduced by which the feet would be turned over as they approached the ground, and be held at the proper angle ready for the "step." Without the aid of a special diagram it would be difficult to explain in detail how this is managed; and it must suffice to say that the chief feature is a friction-clutch worked by the roller of the foot's spoke.

To the onlooker the manner in which the pedrail crawls over obstacles is almost weird. The writer was shown a small working model of a pedrail, propelled along a board covered with bits of cork, wood, etc. The axle of the wheel scarcely moved upwards at all, and had he not actually seen the obstacles he would have been inclined to doubt their existence. An ordinary wheel of equal diameter took the obstructions with a series of bumps and bounds that made the contrast very striking.

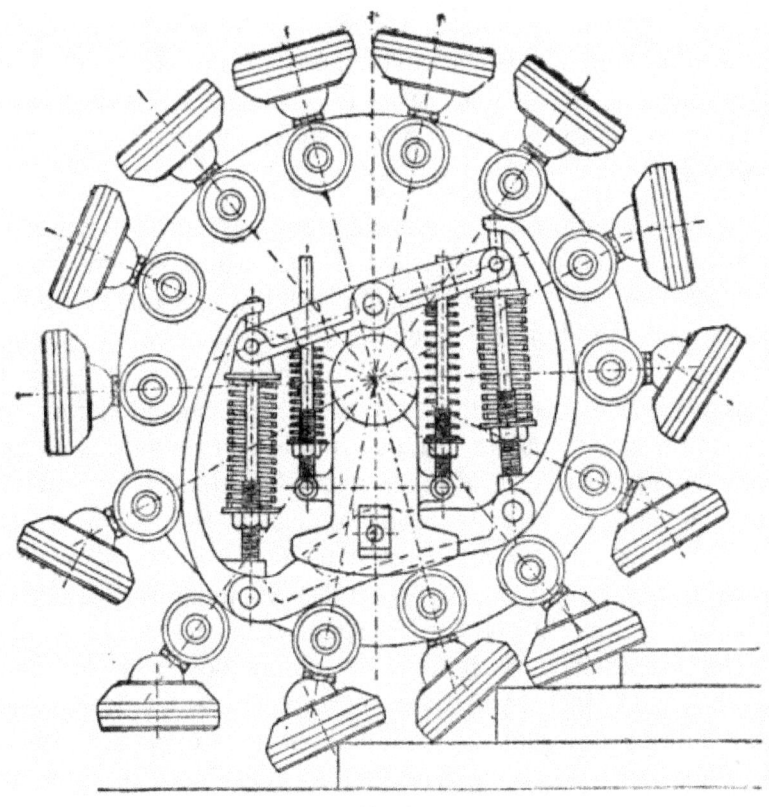

Fig. 4

An extreme instance of the pedrail's capacity would be afforded by the ascent of a flight of steps (see Fig. 4). In such a case the three "peds" carrying the weight of an axle would not be on the same level. That makes no difference, because the frame merely tilts on its top and bottom pivots, the front of the rail rising to a higher level than the back end, and the back spokes being projected by the rail much further than those in front, so that the engine is simply levered over its rollers up an inclined plane. Similarly, in descending, the front spokes are thrust out the furthest, and the reverse action takes place.

With so many moving parts everything must be well lubricated, or the wear would soon become serious. The feet are kept properly greased by being filled with a mixture of blacklead and grease of suitable quality, which requires renewal at long intervals only. The sliding spokes, rollers, and friction-clutches are all lubricated from one central oil-chamber, through a beautiful system of oil-tubes, which provides a circulation of the oil throughout all the moving parts. The central oil-chamber is filled from one orifice, and holds a sufficient supply of oil for a long journey.

We may now turn for a moment from the pedrail itself to the vehicles to which it is attached. Here, again, we are met by novelties, for in his engines Mr. Diplock has so arranged matters, that not only can both front and back pairs of wheels be used as drivers, but both also take part in the steering. As may be imagined, many difficulties had to be surmounted before this innovation was complete. But that it was worth while is evident from the small space in which a double-steering tractor can turn, thanks to both its axles being movable, and from the increased power. Another important feature must also be noticed, viz. that the axles can both tip vertically, so that when the front left wheel is higher than its fellow, the left back wheel may be lower than the right back wheel. In short, *flexibility* and power are the ideals which Mr. Diplock has striven to reach. How far he has been successful may be gathered from the reports of experts. Professor Hele-Shaw, F.R.S., says: "The Pedrail constitutes, in my belief, the successful solution of a walking machine, which, whilst obviating the chief objections to the ordinary wheel running upon the road, can be made to travel anywhere where an ordinary wheel can go, and in many places where it cannot. At the same time it has the mechanical advantages which have made the railway system such a phenomenal success. It constitutes, in my belief, the solution of one of the most difficult mechanical problems, and deserves to be considered as an invention quite apart from any particular means by which it is actuated, whether it is placed upon a self-propelled carriage or a vehicle drawn by any agency, mechanical or otherwise.... The way in which all four wheels are driven simultaneously so as to give the maximum pulling effect by means of elastic connection is in itself sufficient to mark the engine as a most valuable departure from common practice. Hitherto this driving of four wheels has never been successfully achieved, partly because of the difficulty of turning the steering-wheels, and partly because, until the present invention of Mr. Diplock, the front and hind wheels would act against each other, a defect at first experienced and overcome by the inventor in his first engine."

A PEDRAIL TRACTOR ENGAGED IN WAR OFFICE TRIALS

The inventor, Mr. J. B. Diplock, is standing on the left of the group. Observe the manner in which the feet gradually assume a horizontal position as they approach the ground.

On January 8th, 1902, Mr. Diplock tried an engine fitted with two ordinary wheels behind and two pedrails in front. The authority quoted above was present at the trials, and his opinion will therefore be interesting. "The points which struck me immediately were (1) the marvellous ease with which it started into action, (2) the little noise with which it worked.... Another thing which I noticed was the difference in the behaviour of the feet and wheels. The feet did not in any way seem to affect the surface of the road. Throwing down large stones the size of the fist into their path, the feet simply set themselves to an angle in passing over the stones, and did not crush them; whereas, the wheel coming after invariably crushed the stones, and, moreover, distorted the road surface.

"Coming to the top of the hill, I made the Pedrail walk first over 3-inch planks, then 6-inch, and finally over a 9-inch balk....

One could scarcely believe, on witnessing these experiments, that the whole structure was not permanently distorted and strained, whereas it was evidently within the limits of play allowed by the mechanism. As a proof of this the Diplock engine walked down to the works, and I then witnessed its ascent of a lane, beside the engineering works, which had ruts eight or ten inches deep, and was a steep slope. This lane was composed in places of the softest mud, and whereas the wheels squeezed out the ground in all directions, the feet of the Pedrails set themselves at the angles of the rut where it was hard, or walked through the soft and yielding mud without making the slightest disturbance of the surrounding ground.... I came away from that trial with the firm conviction that I had seen what I believe to be the dawn of a new era in mechanical transport."

Mr. Diplock does not regard the pedrail as an end in itself so much as a means to an end, viz. the development of road-borne traffic. For very long distances which must be covered in a minimum of time the railway will hold its own. But there is a growing feeling that unless the railways can be fed by subsidiary methods of transport more effectively than at present, and unless remote country districts, whither it would not pay to carry even a light railway, are brought into closer touch with the busier parts, our communications cannot be considered satisfactory, and we are not getting the best value out of our roads. For many classes of goods *cheapness* of transportation is of more importance than *speed*; witness the fact that coal is so often sent by canal rather than by rail.

Here, then, is the chance for the pedrail tractor and its long train of vehicles fitted with pedrail wheels, which will tend to improve the road surfaces they travel over. Mr. Diplock sets out in his interesting book, *A New System of Heavy Goods Transport on Common Roads*, a scheme for collecting goods from "branch" routes on to "main" routes, where a number of cars will be coupled up and towed by powerful tractors. With ordinary four-wheeled trucks it is difficult to take a number round a sharp corner, since each truck describes a more sudden circle than its predecessor, the last often endeavouring to climb the pavement. Four-wheeled would therefore be replaced by two-wheeled trucks, provided with special couplings to prevent the cars tilting, while allowing them to turn. Cars so connected would follow the same track round a curve.

The body of the car would be removable, and of a standard size. It could be attached to a simple horse frame for transport into the fields. There the farmer would load his produce, and when the body was full it would be returned to the road, picked up by a crane attached to the tractor, swung on to its carriage and wheels, and taken away to join other cars. By making the bodies of such dimensions as to fit three into an ordinary railway truck, they could be entrained easily. On reaching their destination another tractor would lift them out, fit them to wheels, and trundle them off to the consumer. By this method there would be no "breaking bulk" of goods required from the time it was first loaded till it was exposed in the market for sale.

These things are, of course, in the future. Of more present importance is the fact that the War Office has from the first taken great interest in the new invention, which promises to be of value for military transport over ground either rough or boggy. Trials have been made by the authorities with encouraging results. That daring writer, Mr. H. G. Wells, has in his *Land Ironclads* pictured the pedrail taking an offensive part in warfare. Huge steel-plated forts, mounted on pedrails, and full of heavy artillery and machine guns, sweep slowly across the country towards where the enemy has entrenched himself. The forts are impervious alike to shell and bullet, but as they cross ditch or hillock in their gigantic stride, their artillery works havoc among their opponents, who are finally forced to an unconditional surrender.

Even if the pedrail is not made to carry weapons of destruction, we can, after our experiences with horseflesh in the Boer War, understand how important it may become for commissariat purposes. The feats which it has already performed mark it as just the locomotive to tackle the rough country in which baggage trains often find themselves.

To conclude with a more peaceful use for it. When fresh country is opened up, years must often pass before a proper high road can be made, yet there is great need of an organised system of transport. Whither ordinary traction-engines, or carts, even horses, could scarcely penetrate, the pedrail tractor, thanks to its big, flat feet, which give it, as someone has remarked, the appearance of "a cross between a traction-engine and an elephant," will be able to push its way at the forefront of advancing civilisation.

At home we shall have good reason to welcome the pedrail if it frees us from those terrible corrugated tracks so dreaded by the cyclist, and to bless it if it actually beats our roads down into a greater smoothness than they now can boast.

CHAPTER VI

INTERNAL COMBUSTION ENGINES

OIL ENGINES — ENGINES WORKED WITH PRODUCER GAS — BLAST FURNACE GAS ENGINES

IF carbon and oxygen be made to combine chemically, the process is accompanied by the phenomenon called *heat*. If heat be applied to a liquid or gas in a confined space it causes a violent separation of its molecules, and power is developed.

In the case of a steam-engine the fuel is coal (carbon in a more or less pure form), the fluid, water. By burning the fuel under a boiler, a gas is formed which, if confined, rapidly increases the pressure on the walls of the confining vessel. If allowed to pass into a cylinder, the molecules of steam, struggling to get as far as possible from one another, will do useful work on a piston connected by rods to a revolving crank.

We here see the combustion of fuel external to the cylinder, i.e. under the boiler, and the fuel and fluid kept apart out of actual contact. In the gas or oil-vapour engine the fuel is brought into contact with the fluid which does the work, mixed with it, and burnt *inside* the cylinder. Therefore these engines are termed *internal combustion* engines.

Supposing that a little gunpowder were placed in a cylinder, of which the piston had been pushed almost as far in as it would go, and that the powder were fired by electricity. The charcoal would unite with the oxygen contained in the saltpetre and form a large volume of gas. This gas, being heated by the ignition, would instantaneously expand and drive out the piston violently.

A very similar thing happens at each explosion of an internal combustion engine. Into the cylinder is drawn a charge of gas, containing carbon, oxygen, and hydrogen, and also a proportion of air. This charge is squeezed by the inward movement of the piston; its temperature is raised by the compression, and at the proper moment it is ignited. The oxygen and carbon seize on one another and burn (or combine), the heat being increased by the combustion of the hydrogen. The air atoms are expanded by the heat, and work is done on the piston. But the explosion is much gentler than in the case of gunpowder.

During recent years the internal combustion engine has been making rapid progress, ousting steam power from many positions in which it once reigned supreme. We see it propelling vehicles along roads and rails, driving boats through the water, and doing duty in generating stations and smelting works to turn dynamos or drive air-pumps—not to mention the thousand other forms of usefulness which, were they enumerated here, would fill several pages.

A decade ago an internal combustion engine of 100 h.p. was a wonder; to-day single engines are built to develop 3,000 h.p., and in a few years even this enormous capacity will doubtless be increased.

It is interesting to note that the rival systems—gas and steam—were being experimented with at the same time by Robert Street and James Watt respectively. While Watt applied his genius to the useful development of the power latent in boiling water, Street, in 1794, took out letters patent for an engine to be worked by the explosions caused by vaporising spirits of turpentine on a hot metal surface, mixing the vapour with air in a cylinder, exploding the mixture, and using the explosion to move a piston. In his, and subsequent designs, the mixture was pumped in from a separate cylinder under slight pressure. Lenoir, in 1860, conceived the idea of making the piston *suck* in the charge, so abolishing the need of a separate pump; and many engines built under his patents were long in use, though, if judged by modern standards, they were very wasteful of fuel. Two years later Alphonse Beau de Rochas proposed the further improvement of utilising the cylinder, not only as a suction pump, but also as a compressor; since he saw that a compressed mixture would ignite very much more readily than one not under pressure. Rochas held the secret of success in his grasp, but failed to turn it to practical account. The "Otto cycle," invented by Dr. Otto in 1876, is really only Rochas's suggestion materialised. The large majority of internal combustion engines employ this "cycle" of operations, so we may state its exact meaning:—

(1) A mixture of explosive gas and air is drawn into the cylinder by the piston as it passes outwards (*i.e.* in the direction of the crank), through the inlet valve.

(2) The valve closes, and the returning piston compresses the mixture.

(3) The mixture is fired as the piston commences its second journey outwards, and gives the "power" stroke.

(4) The piston, returning again, ejects the exploded mixture through the outlet or exhaust valve, which began to open towards the end of the third stroke.

Briefly stated, the "cycle" is—suction, compression, explosion, expulsion; one impulse being given during each cycle, which occupies two complete revolutions of the fly-wheel. Since the first, second, and third operations all absorb energy, the wheel must be heavy enough to store sufficient momentum during the "power" stroke to carry the piston through all its three other duties.

Year by year, the compression of the mixture has been increased, and improvements have been made in the methods of governing the speed of the engine, so that it may be suitable for work in which the "load" is constantly varying. By doubling, trebling, and quadrupling the cylinders the drive is rendered more and more steady, and the elasticity of a steam-engine more nearly approached.

The internal combustion engine has "arrived" so late because in the earlier part of last century conditions were not favourable to its development. Illuminating gas had not come into general use, and such coal gas as was made was expensive. The great oilfields of America and Russia had not been discovered. But while the proper fuels for this type of motor were absent, coal, the food of the steam-engine, lay ready to hand, and in forms which, though useless for many purposes, could be advantageously burnt under a boiler.

Now the situation has altered. Gas is abundant; and oil of the right sort costs only a few pence a gallon. Inventors and manufacturers have grasped the opportunity. To-day over

3,000,000 h.p. is developed continuously by the internal combustion engine.

Steam would not have met so formidable a rival had not that rival had some great advantages to offer. What are these? Well, first enter a factory driven by steam power, and carefully note what you see. Then visit a large gas- or oil-engine plant. You will conclude that the latter scores on many points. There are no stokers required. No boilers threaten possible explosions. The heat is less. The dust and dirt are less. The space occupied by the engines is less. There is no noisome smoke to be led away through tall and expensive chimneys. If work is stopped for an hour or a day, there are no fires to be banked or drawn— involving waste in either case.

Above all, the gas engine is more efficient, or, if you like to express the same thing in other words, more economical. If you use only one horse-power for one hour a day, it doesn't much matter whether that horse-power-hour costs 4d. or 5d. But in a factory where a thousand horse-power is required all day long, the extra pence make a big total. If, therefore, the proprietor finds that a shilling's-worth of gas or oil does a quarter as much work again as a shilling's-worth of coal, and that either form of fuel is easily obtained, you may be sure that, so far as economy is concerned, he will make up his mind without difficulty as to the class of engine to be employed. A pound of coal burnt under the best type of steam-engine gives but 10 per cent. of its heating value in useful work. A good oil-engine gives 20-25 per cent., and in special types the figures are said to rise to 35-40 per cent. We may notice another point, viz. that, while a steam-engine must be kept as hot as possible to be efficient, an internal combustion engine must be cooled. In the former case no advantage, beyond increased efficiency, results. But in the latter the water passed round the cylinders to take up the surplus heat has a value for warming the building or for manufacturing processes.

Putting one thing with another, experts agree that the explosion engine is the prime mover of the future. Steam has apparently been developed almost to its limit. Its rival is but half-grown, though already a giant.

Some internal combustion engines use petroleum as their fuel, converting it into gas before it is mixed with air to form the charge; others use coal-gas drawn from the lighting mains; "poor gas" made in special plants for power purposes; or natural gas issuing from the ground. Natural gas occurs in very large quantities in the United States, where it is conveyed through pipes under pressure for hundreds of miles, and distributed among factories and houses for driving machinery, heating, and cooking. In England and Europe the petroleum engine and coal-gas engine have been most utilised; but of late the employment of smelting-furnace gases—formerly blown into the air and wasted—and of "producer" gas has come into great favour with manufacturers. The latest development is the "suction" gas engine, which makes its own gas by drawing steam and air through glowing fuel during the suction stroke.

We will consider the various types under separate headings devoted

(1) To the oil-fuel engine,

(2) The producer-gas engine and the suction-gas engine,

(3) Blast-furnace gas engines,

with reference to the installations used in connection with the last two.

All explosion engines (excepting the very small types employed on motor cycles) have a water-jacket round the cylinders to absorb some of the heat of combustion, which would otherwise render the metal so hot as to make proper lubrication impossible, and also would unduly expand the incoming charge of gas and air before compression. The ideal engine would take in a full charge of cold mixture, which would receive no heat from the walls of the cylinder, and during the explosion would pass no heat through the walls. In other words, the ideal metal for the cylinders would be one absolutely non-receptive of heat. In the absence of this, engineers are obliged to make a compromise, and to keep the cylinder at such a temperature that it can be lubricated fittingly, while not becoming so cold as to absorb *too much* of the heat of explosion.

These fall into two main classes:—

(*a*) Those using light, volatile, mineral oils—such as petrol and benzoline—and alcohol, a vegetable product.

(*b*) Those using heavy oils, such as paraffin oil (kerosene) and the denser constituents of rock-oil left in the stills after the kerosene has been driven off. American petroleum is rich in burning-oil and petrol; Russian in the very heavy residue, called *astakti*. Given the proper apparatus for vaporisation, mineral oils of any density can be used in the explosion engine.

The first class is so well known as the mover of motor vehicles and boats that we need not linger here on it. It may, however, be remarked that engines using the easily-vaporised oils are not of large powers, since the fuel is too expensive to make them valuable for installations where large units of power are needed. They have been adopted for locomotives on account of their lightness, and the ease with which they can be started. Petrol vaporises at ordinary temperatures, so that air merely passed over the spirit absorbs sufficient vapour to form an explosive mixture. The "jet" carburetter, now generally employed, makes the mixture more positive by atomising the spirit as it passes through a very fine nozzle into the mixing chamber under the suction from the cylinder. On account of their small size spirit engines work at very high speeds as compared with the large oil or gas engine. Thus, while a 2,000 h.p. Körting gas engine develops full power at eighty-five revolutions a minute, the tiny cycle motor must be driven at 2,000 to 3,000 revolutions. Speaking generally, as the size increases the speed decreases.

Of heavy oil engines there are some dozens of well-tried types. They differ in their methods of effecting the following operations.

1. The feeding of the oil fuel to the engine.

2. The conversion of the oil into vapour.

3. The ignition of the charge.

4. The governing of speed.

All these engines have a vaporiser, or chamber wherein the oil is converted into gas by the action of heat. When starting-up the engine, this chamber must be heated by a specially designed lamp, similar in principle to that used by house painters for burning old paint off wood or metal.

Let us now consider the operations enumerated above in some detail.

1. *The oil supply.* Fuel is transferred from the storage tank to the vaporiser either by the action of gravity through a regulating device to prevent "flooding," or by means of a small pump, or by the suction of the piston, which *lifts* the liquid. In some engines the air and gas enter the cylinder through a single valve; in others through separate valves.

2. *Vaporisation.* As already remarked, the vaporising chamber must be heated to start the engine. When work has begun the lamp may be removed if the engine is so designed that the chamber stores up sufficient heat in its walls from each explosion to vaporise the charge for the next power stroke. The Crossley engine has a lamp continuously burning; the Hornsby-Ackroyd depends upon the storage of heat from explosions in a chamber opening into the cylinder. The best designs are fairly equally divided between the two systems.

3. *Ignition* of the compressed charge is effected in one of four ways: by bringing the charge, at the end of the compression stroke, into contact with a closed tube projecting from the cylinder and heated outside by a continuously burning lamp; by the heat stored in some part of the combustion chamber (*i.e.* that portion of the cylinder not swept by the piston); by an electric spark; or by the mere heat of compression. The second and third methods are confined to comparatively few makes; and the Diesel Oil Engine (of which more presently) has a monopoly of the fourth.

4. *Governing.* All engines which turn machinery doing intermittent work—such as that of a sawmill, or electric generating plant connected with a number of motors—must be very carefully guarded from overrunning. Imagine the effect on an engine which is putting out its whole strength and getting full charges of fuel, if the belt suddenly slipped off and it were

"allowed its head." A burst fly-wheel would be only one of the results. The steam-engine is easily controlled by the centrifugal action of a ball-governor, which, as the speed increases, gradually spreads its balls and lifts a lever connected with a valve in the steam supply pipe. Owing to its elastic nature, steam will do useful work if admitted in small quantities to the cylinder. But a difficulty arises with the internal combustion engine if the *supply* of mixture is similarly throttled, because a loss of quantity means loss of compression and bad ignition. Many oil engines are therefore governed by apparatus which, when the speed exceeds a certain limit, cuts off the supply altogether, either by throwing the oil-pump temporarily out of action, or by lifting the exhaust valve so that the movement of the piston causes no suction—the "hit-and-miss" method, as it is called.

The means adopted depends on the design of the engine; and it must be said that, though all the devices commonly used effect their purpose, none are perfect; this being due rather to the nature of an internal explosion engine than to any lack of ingenuity on the part of inventors. The steadiest running is probably given with the throttle control, which diminishes the supply. On motor cars this method has practically ousted the "hit-and-miss" governed exhaust valve; but in stationary engines we more commonly find the speed controlled by robbing the mixture of the explosive gas in inverse proportion to the amount of the work required from the engine.

THE DIESEL OIL ENGINE,

on account of some features peculiar to it, is treated separately. In 1901 an expert wrote of it that "the engine has not attained any commercial position." Herr Rudolph Diesel, the inventor, has, however, won a high place for his prime-mover among those which consume liquid fuel, on account of its extraordinary economy. The makers claim—as the result of many tests—that with the crude rock-oil (costing in bulk about 2d. a gallon) which it uses, a horse-power can be developed for one hour by this engine for *one-tenth of a penny*. The daily fuel bill for a 100 h.p. engine running ten hours per day would therefore be 8s. 4d. To compete with the Diesel engine a steam installation would have to be of the very highest class of triple-expansion type, of not less than 400 h.p., and using every hour per horse-power only $1\frac{3}{4}$ lbs. of coal at 9s. per ton. Very few large steam-engines work under conditions so favourable, and with small sizes 3-4 lbs. of coal would be burnt for every "horse-power-hour."

The Diesel differs from other internal combustion engines in the following respects:—

1. It works with very much higher compression.

2. The ignition is spontaneous, resulting from the high compression of the charge alone.

3. The fuel is not admitted into the cylinder until the power-stroke begins, and enters in the form of a fine spray.

4. The combustion of the fuel is much slower, and therefore gives a more continuous and elastic push to the piston.

The engine works on the ordinary Otto cycle. To start it, air compressed in a separate vessel is injected into the cylinder. The piston flies out, and on its return squeezes the air to about 500 lbs. to the square inch, thus rendering it incandescent.[11] Just as the piston begins to move out again a valve in the cylinder-head opens, and a jet of pulverised oil is squirted in by air compressed to 100 lbs. per square inch more than the pressure in the cylinder. The vapour, meeting the hot air, burns, but comparatively slowly: the pressure in the cylinder during the stroke decreasing much more gradually than in other engines. Governing is effected by regulation of the amount of oil admitted into the cylinder.

In spite of its high compression this engine runs with very little vibration. The writer saw a penny stand unmoved on its edge on the top of a cylinder in which the piston was reciprocating 500 times a minute!

ENGINES WORKED BY PRODUCER-GAS

These engines are worked by a special gas generated in an apparatus called a "producer." If air is forced through incandescent carbon in a closed furnace its oxygen unites with the carbon and forms carbonic acid gas, known chemically as CO_2, because every molecule of the gas contains one atom of carbon and two of oxygen. This gas, being the product of combustion, cannot burn (*i.e.* combine with more oxygen), but as it passes up through the glowing coke, coal, or other fuel, it absorbs another carbon atom into every molecule, and we have C_2O_2, or 2 CO, which we know as *carbon monoxide*. This gas may be seen burning on the top of an open fire with a very pale blue flame, as it once more combines with oxygen to form carbonic acid gas.

The carbon monoxide is valuable as a heating agent, and when mixed with air forms an explosive mixture.

If along with the air sent into our furnace there goes a proportion of steam, further chemical action results. The oxygen of the steam combines with carbon to form carbon monoxide, and sets free the hydrogen. The latter gas, when it combines with oxygen in combustion, causes intense heat; so that if from the furnace we can draw off carbon monoxide and hydrogen, we shall be able to get a mixture which during combustion will set up great heat in the cylinder of an engine.

In 1878 Mr. Emerson Dowson invented an apparatus for manufacturing a gas suitable for power plant, the gas being known as Producer or Poor Gas, the last term referring to its poorness in hydrogen as compared with coal and other gases. While the hydrogen is a desirable ingredient in an explosive charge, it must not form a large proportion, since under compression it renders the mixture in which it takes part dangerously combustible, and liable to spontaneous ignition before the piston has finished the compression stroke. Water-gas, very rich in hydrogen, and made by a very similar process, is therefore not suitable for internal combustion engines.

There are many types of producers, but they fall under two main heads, *i.e.* the *pressure* and the *suction*.

The *pressure* producer contains the following essential parts:
—

The *generator*, a vertical furnace fed from the top through an air-tight trap, and shut off below from the outside atmosphere by having its foot immersed in water. Any fuel or ashes which fall through the bars into the water can be abstracted without spoiling the draught. Air and steam are forced into the generator, and pass up through the fuel with the chemical results already described. The gases then flow into a *cooler*, enclosed in a water-jacket, through which water circulates, and on into a *scrubber*, where they must find their way upwards through coke kept dripping with water from overhead jets. The water collects impurities of all sorts, and the gas is then ready for storage in the gas-holders or for immediate use in the engines.

A pound of anthracite coal thus burnt will yield enough gas to develop 1 h.p. for one hour.

Suction Gas Plants.—With these gas is not stored in larger quantities than are needed for the immediate work of the engine. In fact, the engine itself during its suction strokes *draws* air and steam through a very small furnace, coolers, and scrubbers direct into the cylinder. The furnace is therefore fed with air and water, not by pressure from outside, but by suction from inside, hence the name "suction producer." At the present time suction gas engines are being built for use on ships, since a pound of fuel thus consumed will drive a vessel further than if burnt under a steam boiler. Very possibly the big ocean liners of twenty years hence may be fitted with such engines in the place of the triple and quadruple expansion steam machinery now doing the work.

BLAST-FURNACE GAS ENGINES

Every iron blast-furnace is very similar in construction and action to the generator of a producer-gas plant. Into it are fed through a hopper, situated in the top, layers of ore, coal or coke, and limestone. At the bottom enters a blast of air heated by passing through a stove of firebrick raised to a high temperature by the carbon monoxide gas coming off from the furnace. When the stove has been well heated the gas supply is shut off from it and switched to the engine-house to create power for driving the huge blowers.

The gas contains practically no hydrogen, as the air sent through the furnace is dry; but since it will stand high compression, it is very suitable for use in large engines. Formerly all the gas from the furnace was expelled into the open air and absolutely wasted; then it was utilised to heat the forced draught to the furnace; next, to burn under boilers; and last of all, at the suggestion of Mr. B. H. Thwaite, to operate internal combustion engines for blowing purposes. Thus, in the fitness of things, we now see the biggest gas engines in the world installed where gas is created in the largest quantities, and an interesting cycle of actions results. The engine pumps the air; the air blows the furnace and melts the iron out of the ore; the furnace creates the gas; the gas heats the air or works the engines to pump more air. So engines and furnace mutually help each other, instead of all the obligation being on the one side.

When, a few years ago, the method was first introduced, engines were damaged by the presence of dust carried with the gas from the furnace. Mr. B. H. Thwaite has, however, perfected means for the separation of injurious matter, and blast-furnace gas is coming into general use in England and on the Continent. Some idea of the power which has been going to waste in ironworks for decades past may be gathered from a report of Professor Hubert after experiments made in 1900. He says that engines of large size do not use more than 100 cubic feet of average blast-furnace gas per effective horse-power-hour, which is less than one-fourth of the consumption of gas required to develop the same power from boilers and good modern condensing steam-engines, so that there is an immense surplus of power to be obtained from a blast-furnace if the blowing engines are worked by the gas it generates, a surplus which can be still further increased if the gas is properly cleaned. It is estimated that for every 100 tons of coke used in an ordinary Cleveland blast-furnace, after making ample allowance for gas for the stoves and power for the lifts, pumps, etc., and for gas for working the necessary blowing engines, there is a surplus of at least *1,500 h.p.*; so that by economising gas by cleaning, and developing the necessary power by gas engines, every furnace owner would have a very large surplus of power for his steel or other works, or for selling in the form of electricity or otherwise.

Yet all this gas had been formerly turned loose for the breezes to warm their fingers at! Truly, as an observant writer has recorded, the sight of a special plant being put up near a blast furnace to manufacture gas for the blowing engines suggests the pumping of water uphill in order to get water-power!

Messrs. Westgarth and Richardson, of Middlesbrough; the John Cockerill Company, of Seraing, Belgium; and the De la Vergne Company, of New York, are among the chief makers of the largest gas engines in the world, ranging up to 3,750 h.p. each. These immense machines, some with fly-wheels 30 feet in diameter, and cylinders spacious enough for a man to stand erect in, work blowers for furnaces or drive dynamos. At the works of the manufacturers mentioned the engines helped to make the steel, and turn the machinery for the creation of brother monsters.

GIGANTIC GAS ENGINES

Five of sixteen 2,000 h.p. Körting Gas Engines built by the De la Vergne Company of New York City for blowing the blast furnaces of the Lackawanna Steel Company. The gas-engine plant at these works is the largest in the world. Notice the man to the left.

This use of a "bye-product" of industry is remarkable, but it can be paralleled. Furnace slag, once cast away as useless, is now recognised to be a valuable manure, or is converted into bricks, tiles, cement, and other building materials. Again, the former waste from the coal-gas purifier assumes importance as the origin of aniline dyes, creosote, saccharine, ammonia, and oils. We really appear to be within sight of the happy time when waste will be unknown. And it therefore is curious that we still burn gas as an illuminant, when the same, if made to work an engine, would give more lighting power in the shape of electric current supplying incandescent lamps.

FOOTNOTE:

11. The fact that air is heated to combustion point by compression has long been known to the Chinese. In *The River of Golden Sand*, Captain Gill writes: "The natives have an apparatus by which they strike a light by compressed air. The apparatus consists of a wooden cylinder $2\frac{1}{2}$ inches long by $\frac{3}{4}$ inch in diameter. This is closed at one end; the bore being about the size of a stout quill pen, an air-tight piston fits into this with a large flat knob at the top. The other end of the piston is slightly hollowed out, and a very small piece of tinder is placed on the top thus formed. The cylinder is held in one hand, the piston inserted and pushed about half-way down; a very sharp blow is then delivered with the palm of the hand on to the top of the knob; the hand must at the same time close on the knob, and instantly withdraw the piston, when the tinder will be found alight. The compression of the air produces heat enough to light the tinder; but this will go out again unless the piston is withdrawn very sharply. I tried a great many times, but covered myself with confusion in fruitless efforts to get a light, for the natives never miss it."

CHAPTER VII

MOTOR-CARS

THE MOTOR OMNIBUS — RAILWAY MOTOR-CARS

The development of the motor-car has been phenomenal. Early in 1896 the only mechanically moved vehicles to be seen on our roads were the traction-engine, preceded by a man bearing a red flag, the steam-roller, and, in the towns, a few trams. To-day the motor is apparent everywhere, dodging through street traffic, or raising the dust of the country roads and lanes, or lumbering along with its load of merchandise at a steady gait.

As a purely speed machine the motor-car has practically reached its limit. With 100 h.p. or more crowded into a vehicle scaling only a ton, the record rate of travel has approached two miles in a minute on specially prepared and peculiarly suitable tracks. Even up steep hills such a monster will career at nearly eighty miles an hour.

Next to the racing car comes the touring car, engined to give sixty miles an hour on the level in the more powerful types, or a much lower speed in the car intended for quieter travel, and for people who are not prepared to face a big bill for upkeep. The luxury of the age has invaded the design of automobiles till the gorgeously decorated and comfortably furnished Pullman of the railway has found a counterpart in the motor caravan with its accommodation for sleeping and feeding. While the town dweller rolls along in electric landaulet, screened from wind and weather, the tourist may explore the roads of the world well housed and lolling at ease behind the windows of his 2,000-guinea machine, on which the engineer and carriage builder have lavished their utmost skill.

The taunt of unreliability once levelled—and with justice—at the motor-car, is fast losing its force, owing to the vast improvements in design and details which manufacturers have been stimulated to make. The motor-car industry has a great future before it, and the prizes therein are such as to tempt both inventor and engineer. Every week scores of patents are granted for devices which aim at the perfection of some part of a car, its tyres, its wheels, or its engines. Until standard types for all grades of motor vehicles have been established, this restless flow of ideas will continue. Its volume is the most striking proof of the vitality of the industry.

The uses to which the motor vehicle has been put are legion. On railways the motor carriage is catering for local traffic. On the roads the motor omnibus is steadily increasing its numbers. Tradesmen of all sorts, and persons concerned with the distribution of commodities, find that the petrol- or steam-moved car or lorry has very decided advantages over horse traction. Our postal authorities have adopted the motor mail van. The War Office looks to the motor to solve some of its transportation difficulties. In short, the "motor age" has arrived, which will, relatively to the "railway age," play much the same part as that epoch did to the "horse age." At the ultimate effects of the change we can only guess; but we see already, in the great acceleration of travel wherever the motor is employed, that many social institutions are about to be revolutionised. But for the determined opposition in the 'thirties of last century to the steam omnibus we should doubtless live to-day in a very different manner. Our population would be scattered more broadcast over the country instead of being herded in huge towns. Many railways would have remained unbuilt, but our roads would be kept in much better condition, special tracks having been built for the rapid travel of the motor. We have only to look to a country now in course of development to see that the road, which leads everywhere, will, in combination with the motor vehicle, eventually supplant, or at any rate render unnecessary, the costly network of railways which must be a network of very fine mesh to meet the needs of a civilised community.

In the scope of a few pages it is impossible to cover even a tithe of the field occupied by the ubiquitous motor-car, and we must, therefore, restrict ourselves to a glance at the manufacture of its mechanism, and a few short excursions into those developments which promise most to alter our modes of life.

We will begin with a trip over one of the largest motor

factories in the world, selecting that of Messrs. Dion and Bouton, whose names are inseparable from the history of the modern motor-car. They may justly claim that to deal with the origin, rise, and progress of the huge business which they have built up would be to give an account, in its general lines, of all the phases through which the motor, especially the petrol motor, has passed from its crudest shape to its present state of comparative perfection.

The Count Albert de Dion was, in his earlier days, little concerned with things mechanical. He turned rather to the fashionable pursuit of duelling, in which he seems to have made a name. But he was not the man to waste his life in such inanities, and when, one day, he was walking down the Paris boulevards, his attention was riveted by a little clockwork carriage exposed for sale among other New Year's gifts. That moment was fraught with great consequences, for an inventive mind had found a proper scope for its energy. Why, thought he, could not real cars be made to run by some better form of motive power? On inquiring he learnt that a workman named Bouton had produced the car. The Count, therefore, sought the artisan; with whom he worked out the problem which had now become his aim in life. Hence it is that the names "Dion—Bouton" are found on thousands of engines all over the world.

The partners scored their first successes with steam- and petrol-driven tricycles, built in a small workshop in the Avenue Malakoff in Paris. The works were then transferred to Puteaux, which has since developed into the great automobile centre of the world, and after two more changes found a resting-place on the Quai National. Here close upon 3,000 hands are engaged in supplying the world's requirements in motors and cars. Let us enter the huge block of buildings and watch them at work.

The drawing-office is the brain of the factory. Within its walls new ideas are being put into practical shape by skilled draughtsmen. The drawings are sent to the model-making shop, where the parts are first fashioned in wood. The shop contains dozens of big benches, circular saws, and planing machines, one of them in the form of a revolving drum carrying a number of planes, which turns thousands of times a minute, and shapes off the rough surface of the blocks of hard wood as if it were so much clay. These blocks are cut, planed, and turned, and then put into the hands of a remarkably skilled class of workmen, who, with rule, calliper, and chisel, shape out cylinders and other parts to the drawings before them with wonderful patience and exactness.

After the model has been fashioned, the next step is to make a clay mould from the same, with a hole in the top through which the molten metal is poured. The foundry is most picturesque in a lurid, Rembrandtesque fashion: "It is black everywhere. The floor, walls, and roof are black, and the foundry hands look like unwashed penitents in sackcloth and ashes. At the end of the building there is a raised brickwork, and when the visitor is able to see in the darkness, he distinguishes a number of raised lids along the top, while here and there are strewn about huge iron ladles like buckets. On the foreman giving the word, a man steps up on the brickwork and removes the lid, when a column of intense white light strikes upwards. It gives one the impression of coming from the bowels of the earth, like a hole opening out in a volcano. The man bestrides the aperture, down which he drops the ladle at the end of a long pole, and then pulling it up again full of a straw-coloured, shining liquid, so close to him that we shudder at the idea of its spilling over his legs and feet, he pours the molten metal into a big ladle, which is seized by two men who pour the liquid into the moulds. The work is more difficult than it looks, for it requires a lot of practice to fill the moulds in such a way as to avoid blow-holes and flaws that prove such a serious item in foundry practice."

In the case-hardening department, next door, there are six huge ovens with sliding fronts. Therein are set parts which have been forged or machined, and are subjected to a high temperature while covered in charcoal, so that the skin of the metal may absorb carbon at high temperatures and become extremely tough. All shafts, gears, and other moving parts of a car are subjected to this treatment, which permits a considerable reduction in the weight of metal used, and greatly increases its resistance to wear. After being "carbonised," the material is tempered by immersion in water while of a certain heat, judged by the colour of the hot metal.

We now pass to the turning-shop, where the cylinders are bored out by a grinding disc rapidly rotating on an eccentric shaft, which is gradually advanced through the cylinder as it revolves. The utmost accuracy, to the 1/10,000 part of an inch, is necessary in this operation, since the bore must be perfectly cylindrical, and also of a standard size, so that any standard piston may exactly fit it. After being bored, or rather ground, the walls of the cylinder are highly polished, and the article is ready for testing. The workman entrusted with this task hermetically closes the ends by inserting the cylinder between the plates of an hydraulic press, and pumps in water to a required pressure. If there be the slightest crack, crevice, or hole, the water finds its way through, and the piece is condemned to the rubbish heap.

In the "motor-room" are scores of cylinders, crank-cases, and gears ready for finishing. Here the outside of bored cylinders is touched up by files to remove any marks and rough projections left by the moulds. The crank-cases of aluminium are taken in hand by men who chisel the edges where the two halves fit, chipping off the metal with wonderful skill and precision. The edges are then ground smooth, and after the halves have been accurately fitted, the holes for the bolts connecting them are drilled in a special machine, which presents a drill to each hole in succession.

Having seen the various operations which a cylinder has to go through, we pass into another shop given up to long lines of benches where various motor parts are being completed. Each piece, however small, is treated as of the utmost importance, since the failure of even a tiny pin may bring the largest car to a standstill. We see a man testing pump discs against a standard template to prove their absolute accuracy. Close by, another man is finishing a fly-wheel, chipping off specks of metal to make the balance true. We now understand that machine tools cannot utterly displace the human hand and eye. The fitters, with touches of the file, remove matter in such minute quantities that its removal might seem of no consequence. But "matter in the wrong place" is the cause of many breakdowns.

We should naturally expect that engines cast from the same pattern, handled by the same machines, finished by the same men, would give identical results. But as two bicycles of similar make will run differently, so do engines of one type develop peculiarities. The motors are therefore taken into a testing-room and bolted to two rows of benches, forty at a time. Here they run under power for long periods, creating a deafening uproar, until all parts work "sweetly." The power of the engines is tested by harnessing them to dynamos and noting the amount of current developed at a certain speed.

We might linger in the departments where accumulators, sparking plugs, and other parts of the electrical apparatus of a car are made, or in the laboratory where chemists pry into the results of a new alloy, aided by powerful microscopes and marvellously delicate scales. But we will stop only to note the powerful machine which is stretching and crushing metal to ascertain its toughness. No care in experimenting is spared. The chemist, poring over his test tubes, plays as important a part in the construction of a car as the foundry man or the turner.

The machine-shop is an object-lesson among the tools noticed in previous chapters of this book. "Here is a huge planing machine travelling to and fro over a copper bar. A crank shaft has been cut out of solid steel by boring holes close together through a thick plate, and the two sides of the plate have been broken off, leaving the rough shaft with its edges composed of a considerable number of semicircles. The shaft is slowly rotated on a lathe, and tiny clouds of smoke arise as the tool nicks off pieces of metal to reduce the shaft to a circular shape. Other machines, with high-speed tool steel, are finishing gear shafts. Fly-wheels are being turned and worm shafts cut. All these laborious operations are carried out by the machines, each under the control of one man whose mind is intent upon the work, ready to stop the machine or adjust the material as may be required. As a contrast to the heavy machines we will pass to the light automatic tools which are grouped in a gallery.... The eye is bewildered by the moving mass, but the whirling of the pulley shafts and the clicking of the capstan lathes is soothing to the ear, while the mind is greatly impressed by the ingenuity of man in suppressing labour by means of machines, of which half a dozen can be easily looked after by one hand, who has nothing to do but to see that they are fed with material. A rod of steel is put into the machine, and the turret, with half a dozen different tools, presents first one and then the other to the end of the rod bathed in thick oil, so that it is rapidly turned, bored, and shaped into caps, nuts, bolts, and the scores of other little accessories required in fitting up a motor-car. On seeing how all this work is done mechanically and methodically, with scarcely any other expense but the capital required in the upkeep of the machines and in driving them, one wonders how the automobile industry could be carried on without this labour-saving mechanism. In any event, if all these little pieces had to be turned out by hand, it is certain that the cost of the motor-car would be considerably more than it is, even if it did not reach to such a figure as to make it prohibitive to all but wealthy buyers. Down one side of the gallery the machines are engaged in cutting gears with so much precision that, when tested by turning them together on pins on a bench at the end of the gallery, it is very rare indeed that any one of them is found defective. This installation of automatic tools is one of the largest of its kind in a motor-car works, if not in any engineering shop, and each one has been carefully selected in view of its efficiency for particular classes of work, so that we see machines from America, England, France, and Germany."

In the fitting-shops the multitude of parts are assembled to form the *chassis* or mechanical carriage of the car, to which, in a separate shop, is added the body for the accommodation of passengers. The whole is painted and carefully varnished after it has been out on the road for trials to discover any weak spot in its anatomy. Then the car is ready for sale.

When one considers the racketing that a high-powered car has to stand, and the high speed of its moving parts, one can understand why those parts must be made so carefully and precisely, and also how this care must conduce to the expense of the finished article. It has been said that it is easy to make a good watch, but difficult to make a good motor; for though they both require an equal amount of exactitude and skill, the latter has to stand much more wear in proportion. When you look at a first-grade car bearing a great maker's name, you have under your eyes one of the most wonderful pieces of mechanism the world can show.

We will not leave the de Dion-Bouton Works without a further glance at the human element. The company never have a slack time, and consequently can employ the same number of people all the year round. They pride themselves on the fact that the great majority of the men have been in their employ for several years, with the result that they have around them a class of workmen who are steady, reliable and, above all, skilful in the particular work they are engaged upon. There are more than 2,600 men and about 100 women, these latter being employed chiefly in the manufacture of sparking plugs and in other departments where there is no night work. They are mostly the wives or widows of old workmen, and in thus finding employment for them the firm provides for those who would otherwise be left without resource, and at the same time earns the gratitude of their employés.

NOTE.—The author gratefully acknowledges the help given by Messrs. de Dion-Bouton, Ltd., in providing materials for this account of their works.

THE MOTOR OMNIBUS

Prior to the emancipation of the road automobile in 1896, permission had been granted to corporations to run trams driven by mechanical power through towns. The steam tram, its engine protected by a case which hid the machinery from the view of restive horses, panted up and down our streets, drawing one or more vehicles behind it. The electric tram presently came over from America and soon established its superiority to the steamer with respect to speed, freedom from smell and smoke, and noiselessness: the system generally adopted was that invented in 1887 by Frank J. Sprague, in which an overhead cable supported on posts or slung from wires spanning the track carries current to a trolley arm projecting from the vehicle. The return current passes through the rails, which are made electrically continuous by having their individual lengths either welded together or joined by metal strips.

In America, where wide streets and rapidly growing cities are the rule, the electric tramway serves very useful ends; the best proof of its utility being the total mileage of the tracks. Statistics for 1902 show that since 1890 the mileage had increased from 1,261 to 21,920 miles; and the number of passengers carried from 2,023,010,202 to 4,813,466,001, or an increase of 137·94 per cent. It is interesting to note that electricity has in the United States almost completely ousted steam and animal traction so far as street cars are concerned; since the 5,661 miles once served by animal power have dwindled to 259, and steam can claim only 169 miles of track.

Next to the United States comes Germany as a user of electricity for tractive purposes; though she is a very bad second with only about 6,000 miles of track; and England takes third place with about 3,000 miles. That the British Isles, so well provided with railways, should be so poorly equipped with tramways is comprehensible when we consider the narrowness of the streets of her largest towns, where a good service of public vehicles is most needed. The installation of a tram-line necessitates the tearing up of a street, and in many cases the closing of that street to traffic. We can hardly imagine the dislocation of business that would result from such a blockage of, say, the Strand and High Holborn; but since it has been calculated that no less than five millions of pounds sterling are lost to our great metropolis yearly by the obstructions of gas, water, telegraph, and telephone operations, which only partially close a thoroughfare, or by the relaying of the road surface, which is not a very lengthy matter if properly conducted, we might reckon the financial loss resulting from the laying of tram-rails at many millions.

Even were they laid, the trouble would not cease, for a tram is confined to its track, and cannot make way for other traffic. This inadaptability has been the cause of the great outcry lately raised against the way in which tram-line companies have monopolised the main streets and approaches to many of our largest towns. While the electric tram is beneficial to a large class of people, as a cheap method of locomotion between home and business, it sadly handicaps all owners of vehicles vexatiously delayed by the tram. At Brentford, to take a notorious example, the double tram-line so completely fills the High Street that it is at places impossible for a cart or carriage to remain at the kerbstone.

Another charge levelled with justice at the tram-line is that the rails and their setting are dangerous to cyclists, motorists, and even heavy vehicles, especially in wet weather, when the "side-slip" demon becomes a real terror.

English municipalities are therefore faced by a serious problem. Improved locomotion is necessary; how can it best be provided? By smooth-running, luxurious, well-lighted electric trams, travelling over a track laid at great expense, and a continual nuisance to a large section of the community; or by vehicles independent of a central source of power, and free to move in any direction according to the needs of the traffic? Where tramways exist, those responsible for laying them at the rate of several thousand pounds per mile are naturally reluctant to abandon them. But where the fixed track has not yet arrived an alternative method of transport is open, viz. the automobile omnibus. Quite recently we have seen in London and other towns a great increase in the number of motor buses, which often ply far out into the country. From the point of speed they are very superior to the horsed vehicle, and statistics show that they are also less costly to run in proportion to the fares carried, while passengers will unanimously acknowledge their greater comfort. To change from the ancient, rattling two-horse conveyance, which jolts us on rough roads, and occasionally sends a thrill up the spine when the brakes are applied, to the roomy steam- or petrol-driven bus, which overtakes and threads its way through the slower traffic, is a pleasant experience. So the motor buses are crowded, while the horsed rivals on the same route trundle along half empty. Since the one class of vehicles can travel at an average pace of ten miles an hour, as against the four miles an hour of the other, no wonder that this should be so. Even if the running costs of a motor bus for a given distance exceed that of an electric tram, we must remember that, whereas a bus runs on already existing roads, an immense amount of capital must be sunk in laying the track for the tram, and the interest on this sum has to be added to the total running costs.

The next decade will probably decide whether automobiles or trams are to serve the needs of the community in districts where at present no efficient service of any kind exists. In London motor buses are being placed on the roads by scores, and the day cannot be far distant when the horse will disappear from the bus as it is already fast vanishing from the front of the tram.

Both petrol and steam, and in some cases a combination of petrol and electricity, are used to propel the motor bus. It has not yet been decided which form of power yields the best results. Petrol is probably the cheaper fuel, but steam gives the quieter running; and could electric storage batteries be made sufficiently light and durable they would have a strong claim to precedence. There has lately appeared a new form of accumulator—the von Rothmund—which promises well, since weight for weight it far exceeds in capacity any other type, and is so constructed that it will stand a lot of rough usage. A car fitted with a von Rothmund battery scaling about 1,500 lbs. has run 200 miles on one charge, and it is anticipated that with improvements in motors a 1,100-lb. battery will readily be run 150 miles as against the 50 miles in the case of a lead battery of equal weight.

There is a large sphere open to the motor bus outside districts where the electric tram would enter into serious competition with it. We have before us a sketch-map of the Great Western Railway, one of the most enterprising systems with regard to its use of motors to feed its rails. No less than thirty road services are in operation, and their number is being steadily augmented. In fact, it looks as if in the near future the motor service will largely supplant the branch railway, blessed with very few trains a day. A motor bus service plying every half-hour between a town and the nearest important main-line station would be more valuable to the inhabitants than half a dozen trains a day, especially if the passenger vehicles were supplemented by lorries for the carriage of luggage and heavy goods.

In this connection we may notice an invention of M. Renard—a motor train of several vehicles towed by a single engine. We have all seen the traction-engine puffing along with its tail of trucks, and been impressed by the weight of the locomotive, and also by the manner in which the train occupies a road when passing a corner. The weight is necessary to give sufficient grip to move the whole train, while the spreading of the vehicles across the thoroughfare on a curve arises from the fact that each vehicle does not follow the path of that preceding it, but describes part of a smaller circle.

M. Renard has, in his motor train, evaded the need for a heavy tractor by providing *every* vehicle with a pair of driving wheels, and transmitting the power to those wheels by a special flexible propeller shaft which passes from the powerful motor on the leading vehicle under all the other vehicles, engaging in succession with mechanism attached to all the driving axles. In this manner each car yields its quotum of adhesion for its own propulsion, and the necessity for great weight is obviated. Special couplings ensure that the path taken by the tractor shall be faithfully followed by all its followers. A motor train of this description has travelled from Paris to Berlin and drawn to itself a great deal of attention.

"Will it," asks a writer in *The World's Work*, "ultimately displace the conventional traction-engine and its heavy trailing waggons? Every municipality and County Council is only too painfully cognisant of the dire effects upon the roads exercised by the cumbrous wheels of these unwieldy locomotives and trains. With the Renard train, however, the trailing coaches can be of light construction, carried on ordinary wheels which do not cut up or otherwise damage the roadway surface. Many other advantages inherent in such a train might be enumerated. The most important, however, are the flexibility of the whole train; its complete control; faster speed without any attendant danger; its remarkable braking arrangements as afforded by the continuous propeller shaft gearing directly with the driving-wheels of each carriage; its low cost of maintenance, serviceability, and instant use; and the reduction in the number of men requisite for the attention of the train while on a journey."

Were the system a success, it would find plenty of scope to convey passengers and commodities through districts too sparsely populated to render a railway profitable. People would talk about travelling or sending goods by the "ten-thirty motor train," just as now we speak of the "eleven-fifteen to town."

As a carrier and distributer of mails, the motor van has already established a position. To quote but a couple of instances, there are the services between London and Brighton, and Liverpool and Manchester. In the Isle of Wight motor omnibuses connect all the principal towns and villages. Each bus is a travelling post-office in which, by an arrangement with the Postmaster-General, anybody may post letters at the recognised stopping-places or whenever the vehicle has halted for any purpose.

In Paris, London, Berlin, the motor mail van is a common sight. It has even penetrated the interior of India, where the Maharajah of Gwalior uses a specially fitted steam car for the delivery of his private mails. And, as though to show that man alone shall not profit by the new mode of locomotion, Paris owns a motor-car which conveys lost dogs from the different police-stations to the Dogs' Home! In fact, there seems to be no purpose to which a horse-drawn vehicle can be put, which either has not been, or shortly will be, invaded by the motor.

RAILWAY MOTOR-CARS

In the early days of railway construction vehicles were used which combined a steam locomotive with an ordinary passenger carriage. After being abandoned for many years, the "steam carriage" was revived, in 1902, by the London and South Western and Great Western railways for local service and the handling of passenger traffic on branch lines. Since that year rail motor-cars have multiplied; some being run by steam, others by petrol engines, and others, again, by electricity generated by petrol engines. The first class we need not describe in any detail, as it presents no features of peculiar interest.

The North Eastern has had in use two rail-motors, each fifty-two feet long, with a compartment at each end for the driver, and a central saloon to carry fifty-two passengers. An 80 h.p. four-cylindered Wolseley petrol motor drives a Westinghouse electric generator, which sends current into a couple of 55 h.p. electric motors geared to the running-wheels. An air compressor fitted to the rear bogie supplies the Westinghouse air brakes, while in addition a powerful electric brake is fitted, acting on the rails as well as the wheels. The coach scales thirty-five tons.

The chief advantage of this "composite" system of power transmission is that the engine is kept running at a constant speed, while the power it develops at the electric motors is regulated by switches which control the action of the armature and field magnets. When heavy work must be done the engine is supplied with more gaseous mixture, and the generators are so operated as to develop full power. In this manner all the variable speed gears and clutches necessary when the petrol motor is connected to the driving-wheels are done away with.

The latter system gives, however, greater economy of fuel, and the Great Northern Railway has adopted it in preference to the petrol-electric. This railway has many small branch lines running through thinly populated districts, which, though important as feeders of the main tracks, are often worked at a loss. A satisfactory type of automobile carriage would not only avoid this loss, but also largely prevent the competition of road motors.

The car should be powerful enough to draw an extra van or two on occasion, since horses and heavy luggage may sometimes accompany the passengers. Messrs. Dick, Kerr, and Company have built a car, which, when loaded with its complement of passengers, weighs about sixteen tons. The motive power is supplied by two four-cylinder petrol engines of the Daimler type, each giving 36 h.p. These are suspended on a special frame, independent of that which carries the coach body, so that the passengers are not troubled by the vibration of the engines, even when the vehicle is at rest. The great feature of the car is the lightness of the machinery—only two tons in weight—though it develops sufficient power to move the carriage at fifty miles per hour. After travelling 2,000 miles the machinery showed no appreciable signs of wear; so that the company considers that it has found a reliable type of motor for the working of the short line between Hatfield and Hertford.

Since one man can drive a petrol car, while two—a driver and a stoker—are necessary on a steam car, a considerable reduction in wages will result from the employment of these vehicles.

Engineers find motor-trolleys very convenient for inspecting the lines under their care. On the London and South Western

Railway a trolley driven by a 6–8 h.p. engine, and provided with a change-gear giving six, fifteen, and thirty miles per hour in either direction, is at work. It seats four persons. In the colonies, notably in South Africa, where coal and wood fuel is scarce or expensive, the motor-trolley, capable of carrying petrol for 300 miles' travel, is rapidly gaining ground among railway inspectors.

Makers are turning their attention to petrol shunting engines, useful in goods yards, mines, sewerage works. Firms such as Messrs. Maudslay and Company, of Coventry; the Wolseley Tool and Motor Car Company; Messrs. Panhard and Levassor; Messrs. Kerr, Stuart, and Company have brought out locomotives of this kind which will draw loads up to sixty tons. The fact that a petrol engine is ready for work at a moment's notice, and when idle is not "eating its head off," and has no furnace or boiler to require attention, is very much in its favour where comparatively light loads have to be hauled.

CHAPTER VIII

THE MOTOR AFLOAT

PLEASURE BOATS — MOTOR LIFEBOATS — MOTOR FISHING BOATS — A MOTOR FIRE FLOAT — THE MECHANISM OF THE MOTOR BOAT — THE TWO-STROKE MOTOR — MOTOR BOATS FOR THE NAVY

HAVING made such conquests on land, and rendered possible aerial feats which could scarcely have been performed by steam, the explosion motor further vindicates its versatility by its fine exploits in the water.

At the Paris Exhibition of 1889 Gottlieb Daimler, the inventor who made the petrol engine commercially valuable as an aid to locomotion, showed a small gas-driven boat, which by most visitors to the Exhibition was mistaken for an ordinary steam launch, and attracted little interest. Not deterred by this want of appreciation, Mr. Daimler continued to perfect the idea for which, with a prophet's eye, he saw great possibilities; and soon motor launches became a fairly common sight on German rivers. They were received with some enthusiasm in the United States, as being particularly suitable for the inland lakes and waterways with which that country is so abundantly blessed; but met with small recognition from the English, who might reasonably have been expected to take great interest in any new nautical invention. Now, however, English manufacturers have awaked fully to their error; and on all sides we see boats built by firms competing for the lead in an industry which in a few years' time may reach colossal proportions.

A MODERN CAR AND BOAT

In the background is the racing motor boat "Napier II.", which on a trial trip travelled over the "measured mile" at 30·93 miles per hour. In the foreground is a "Napier" racing car, which has attained a speed of 104·8 miles per hour.

Until quite recently the marine motor was a small affair, developing only a few horse-power. But because the gas-engine for automobile work had been so vastly improved in the last decade, it attracted notice as a rival to steam for driving launches and pleasure boats, and soon asserted itself as a reliable mover of vessels of considerable size. To promote the development of the industry, to test the endurance of the machine, and to show the weak spots of mechanical design, trials and races were organised on much the same lines as those which have kept the motor-car so prominently before the public—races in the Solent, across the Channel, and across the Mediterranean. The speed, as in the case of cars, has risen very rapidly with the motor boat. When, in February, 1905, a Napier racer did some trial spins over the measured mile in the Thames at Long Reach, she attained 28·57 miles per hour on the first run. On turning, the tide was favourable, and the figures rose to 30·93 m.p.h., while the third improved on this by over a mile. Her mean speed was 29·925 m.p.h., or about $\frac{2}{3}$ m.p.h. better than the previous record—standing to the credit of the American *Challenger*. The latter had, however, the still waters of a lake for her venue, so that the Napier's performance was actually even more creditable than the mere figures would seem to imply. At a luncheon which concluded the trial, Mr. Yarrow, who had built the steel hull, said: "To give an idea of what an advance the adoption of the internal combustion engine really represents, I should like to state that, if we were asked to guarantee the best speed we could with a boat of the size of Napier II., fitted with the latest form of steam machinery of as reliable a character as the internal combustion engine in the present boat, we should not like to name more than sixteen knots. So that it may be taken that the adoption of the internal combustion engine, in place of the steam-engine, for a vessel of this size, really represents an additional speed of ten knots an hour. I should here point out that the speed of a vessel increases rapidly with its size. For example: in what is termed a second-class torpedo boat, sixty feet in length, the best speed we could obtain would be twenty knots; but for a vessel of, say, 200 feet in length, with similar but proportionately larger machinery, a speed of thirty knots could be obtained. Therefore, the obtaining of a speed of practically twenty-six knots in the Yarrow-Napier boat, only forty feet in length, points to the possibility, in the not far-distant future, of propelling a vessel 220 feet in length at even forty-five knots per hour. All that remains to be done is to perfect the internal combustion engine, so as to enable large sizes to be successfully made."

Boats of 300 h.p. and upwards are being built; and the project has been mooted of holding a transatlantic race, open to motor boats of all sizes, which should be quite self-contained and able to carry sufficient fuel to make the passage without taking in fresh supplies. In view of the perils that would be risked by all but large craft, and in consideration of the prejudice that motor boats might incur in event of any fatalities, the Automobile Club of France set its face against the venture, and it fell through. It is possible, however, that the scheme may be revived as soon as larger motor boats are afloat, since the Atlantic has actually been crossed by a craft of 12 h.p., measuring only forty feet at the water-line. This happened in 1902, when Captain Newman and his son, a boy twelve years old, started from New York, and made Falmouth Harbour after thirty days of anxious travel over the uncertain and sometimes tempestuous ocean. The boat, named the *Abiel Abbot Low*, carried auxiliary sails of small size, and was not by any means built for such a voyage. The engine—a two-cylinder—burned kerosene. Captain Newman received £1,000 from the New York Kerosene Oil Engine Company for his feat. The money was well earned. Though provided with proper navigating instruments—which he knew how to use well—Newman had a hard time of it to keep his craft afloat, his watches sometimes lasting two days on end when the weather was bad. Yet the brave pair won through; and probably even more welcome than the sense of success achieved and the reward gained was the long two-days' sleep which they were able to get on reaching Falmouth Harbour.

PLEASURE BOATS

We may now consider the pleasure and commercial uses of the motor boat and marine motor. As a means of recreation a small dinghy driven by a low-powered engine offers great possibilities. Its cost is low, its upkeep small, and its handiness very great. Already a number of such craft are furrowing the surface of the Thames, Seine, Rhine, and many other rivers in Europe and America. While racing craft are for the wealthy alone, many individuals of the class known as "the man of moderate means" do not mind putting down £70 to £100 for a neat boat, the maintenance of which is not nearly so serious a matter as that of a small car. Tyre troubles have no counterpart afloat. The marine motor dispenses with change gears. Water being a much more yielding medium than Mother Earth, the shocks of starting and stopping are not such as to strain machinery. Then again, the cooling of the cylinders is a simple matter with an unlimited amount of water almost washing the engine. And as the surface of water does not run uphill, a small motor will show to better advantage on a river than on a road. Thus, a 5 h.p. car will not conveniently carry more than two people if it is expected to climb slopes at more than a crawl. Affix a motor of equal power to a boat which accommodates half a dozen persons, and it will move them all along at a smart pace as compared with the rate of travel given by oars. After all, on a river one does not want to travel fast—rather to avoid the hard labour which rowing undoubtedly does become with a craft roomy enough to be comfortable for a party.

The marine motor also scores under the heading of adaptability. A wagonette could not be converted into a motor-car with any success. But a good-sized row-boat may easily blossom out as a useful self-propelled boat. You may buy complete apparatus—motor, tanks, screw, batteries, etc.—for clamping direct on to the stern, and there you are—a motor boat while you wait! Even more sudden still is the conversion effected by the Motogodille, which may be described as a motor screw and rudder in one. The makers are the Buchet Company, a well-known French firm. "Engine and carburetter, petrol tank, coil, accumulator, lubricating oil reservoir, exhaust box, propeller shaft, and propeller with guard are all provided, so that the outfit requires no additional accessories. For mounting in position at the stern of the boat, the complete set is balanced on a standard, and carries a steering arm, on which the tanks are mounted; and also the stern tube and propeller guard, which are in one solid piece, in addition to the engine. In order that no balancing feats shall be required of the person in charge, there is, on the supporting standard, a quadrant, in the notches of which a lever on the engine frame engages, thus allowing the rigid framework, and therefore the propeller shaft, to be maintained at any angle to the vertical without trouble."[12]

The 2 h.p. engine drives a boat 16 feet long by 4 feet 6 inches beam at $6\frac{1}{2}$ miles per hour through still water. As the Motogodille can be swerved to right or left on its standard, it acts as a very efficient rudder, while its action takes no way off the boat.

For people who like an easy life on hot summer days, reclining on soft cushions, and peeping up through the branches which overhang picturesque streams, there is the motor punt, which can move in water so shallow that it would strand even a row-boat. The Oxford undergraduate of to-morrow will explore the leafy recesses of the "Cher," not with the long pole laboriously raised and pushed aft, but by the power of a snug little motor throbbing gently at the stern. And on the open river we shall see the steam launch replaced by craft having much better accommodation for passengers, while free from the dirt and smells which are inseparable from the use of steam-power. The petrol launch will rival the electric in spaciousness, and the steamer in its speed and power, size for size.

Some people have an antipathy to this new form of river locomotion on account of the risks which accompany the presence of petrol. Were a motor launch to ignite in, say, Boulter's Lock on a summer Sunday, or at the Henley Regatta, there might indeed be a catastrophe. The same danger has before now been flaunted in the face of the automobilist on land; yet cases of the accidental ignition of cars are very, very rare, and on the water would be more rare still, because the tanks can be more easily examined for leaks. Still, it behoves every owner of a launch to keep his eye very widely open for leakage, because any escaping liquid would create a collection of gas in the bottom of the boat, from which it could not escape like the gas forming from drops spilled on the road.

Photo Branger & Cie, Paris.
THE MOTOGODILLE

The Motogodille, or Motor Rudder, consists of a screw propeller fitted to a small Buchet motor. The whole apparatus is mounted on a standard in the stern, and the operator, by moving the inboard arm to right or left, can steer the boat as he wishes. A 2-h.p. motor gives a speed of 5 to 6 miles an hour.

The future popularity of the motor boat is assured. The waterside dweller will find it invaluable as a means of carrying him to other parts of the stream. The "longshoreman" will be able to venture much further out to sea than he could while he depended on muscles or wind alone, and with much greater certainty of returning up to time. A whole network of waterways intersects civilised countries—often far better kept than the roads—offering fresh fields for the tourist to conquer. River scenery and beautiful scenery more often than not go together. The car or cycle may be able to follow the course of a stream from source to mouth; yet this is the exception rather than the rule. We shoot *over* the stream in the train or on our machines; note that it looks picturesque; wonder vaguely whither it flows and whence it comes; and continue our journey, recking little of the charming sights to be seen by anyone who would trust himself to the water.

Hitherto the great difficulty has been one of locomotion. In a narrow stream sailing is generally out of the question; haulage by man or beast becomes tedious, even if possible; and rowing day after day presupposes a good physical condition. In the motor boat the holiday maker has an ideal craft. It occupies little room; can carry fuel sufficient for long distances; is unwearying; and is economical as regards its running expenses. We ought not to be surprised, therefore, if in a few years the jaded business man turns as naturally to a spin or trip on the rivers and canals of his country as he now turns to his car and a rush over the dusty highway. Then will begin another era for the disused canal, the vegetation-choked stream; and our maps will pay more attention to the paths which Nature has water-worn in the course of the ages.

To the scientific explorer also the motor affords valuable help. Many countries, in which roads are practically non-existent, can boast fine rivers fed by innumerable streams. What fields of adventure, sport, and science would be open to the possessor of a fast launch on the Amazon, the Congo, the Mackenzie, or the Orinoco, provided only that he could occasionally replenish his fuel tanks!

MOTOR LIFEBOATS

Turning to the more serious side of life, we find the marine motor still much in evidence. On account of its comparatively short existence it is at present only in the experimental stage in many applications, and time must pass before its position is fully established. Take, for instance, the motor lifeboat lately built for the Royal National Lifeboat Institution. Here are encountered difficulties of a kind very different from those of a racing craft. A lifeboat is most valuable in rough weather, which means more or less water often coming aboard. If the water reached the machinery, troubles with the electrical ignition apparatus would result. So the motor must be enclosed in a water-tight compartment. And if so enclosed it must be specially reliable. Also, since a lifeboat sometimes upsets, the machinery needs to be so disposed as not to interfere with her self-righting qualities. The list might easily be extended.

An account of the first motor life-saver will interest readers, so we once again have recourse to the chief authority on such topics —the *Motor Boat*—for particulars. The boat selected for experiment was an old one formerly stationed at Folkestone, measuring thirty-eight feet long by eight feet beam, pulling twelve oars, double-banked, and of the usual self-righting type, rigged with jib, fore-lug, and mizzen. After she had been hauled up in Mr. Guy's yard, where some of the air-cases under the deck amidships were taken out, a strong mahogany case, measuring four feet long by three feet wide and as high as the gunwales, lined with sheet copper so as to be water-tight, with a close-fitting lid which could be easily removed on shore, was fitted in place, and the whole of the vital parts of the machinery, comprising a two-cylinder motor of 10 h.p., together with all the necessary pumps, carburetter, electric equipment, etc., were fitted inside this case. The engine drives a three-bladed propeller through a long shaft with a disconnecting clutch between, so that for starting or stopping temporarily the screw can be disconnected from the engine. The petrol, which serves as fuel for the engine, is carried in a metal tank stored away inside the forward "end" box, where it is beyond any possibility of accidental damage. Sufficient fuel for a continuous run of over ten hours is carried. The engine is started by a handle fitted on the fore side of the case, which can be worked by two men. The position and size of the engine-case is such that only two oars are interfered with, but it does not follow that the propelling power of the two displaced men is entirely lost, because they can double bank some of the other oars when necessary.

Fitted thus, the lifeboat was tested in all sorts of weather during the month of April, and it was found that she could be driven fairly well against a sea by means of the motor alone; but when it was used to assist the sails the true use of the motor as an auxiliary became apparent, and the boat would work to windward in a way previously unattainable. Neither the pitching or rolling in a seaway, in any weather then obtainable, interfered at all with the proper working or starting of the motor, which worked steadily and well throughout. Having been through these preliminary tests, she was more severely tried. Running over the measured mile with full crew and stores on board, she developed over six knots an hour. The men were then replaced by equivalent weights lashed to the thwarts, and she was capsized by a crane four times, her sails set and the sheets made fast, yet she righted herself without difficulty. An interesting feature of the capsize was that the motor stopped automatically when the boat had partly turned over. This arrangement prevents her from running away from the crew if they should be pitched out. The motor started again after a few turns of the handle, so proving that the protecting compartment had kept the water at bay.

From this account it is obvious that a valuable aid to life-saving at sea has been found. The steam lifeboat, propelled by a jet of water squirted out by pumps below the water line, is satisfactory so long as the boat keeps upright. But in event of an upset the fires must necessarily be extinguished. No such disability attends the petrol-driven craft, and we shall be glad to think that the brave fellows who risk their lives in the cause of humanity will be spared the intense physical toil which a long row to windward in a heavy sea entails. The general adoption of this new ally will take time, and must depend largely on the liberality of subscribers to the fine institution responsible for lifeboat maintenance; but it is satisfactory to learn that the Committee has given the boat in question a practical chance in the open sea by stationing her at Newhaven, Sussex, as a unit in the lifeboat fleet.

MOTOR FISHING BOATS

It is a pretty sight to watch a fishing fleet enter the harbour with its catch, taken far away on the waters beyond the horizon while landsmen slept. The sails, some white, some brown, some wondrously patched and bearing the visible marks of many a hard fight with the wind, belly out in graceful lines as the boats slip past the harbour entrance. No wonder that the painter has so often found subjects for his canvas and brushes among the toilers of the deep.

But underlying the romance and picturesqueness of the craft there is stern business. Those boats may be returning with full cargoes, such as will yield good profits to owner and crew; or, on the other hand, the hold may be empty, and many honest hearts be heavy at the thought of wasted days. A few years ago the Yarmouth herring fleet is said to have returned on one occasion with but a single fish to the credit of the whole fleet! This might have been a mere figure of speech; it stands, at any rate, for many thousands of pounds lost by the hardy fishermen.

When the boats have been made fast, the fish, if already disentangled from the nets, is usually sold at once by auction, the price depending largely on the individual size and freshness of the "catch." Now, with the increase in the number of boats and from other causes, the waters near home have been so well fished over that much longer journeys must be made to the "grounds" than were formerly necessary. Trawling, that is, dragging a large bag-net—its mouth kept open by a beam and weights—along the bottom of the sea for flatfish, has long been performed by powerful steam vessels, which may any day be seen leaving or entering Hull or Grimsby in large numbers. Surface fishing, wherein a long drift-net, weighted at its lower edge and buoyed at the upper edge to enable it to keep a perpendicular position, is used for herring and mackerel, and in this industry wind power alone is generally used by British fishermen.

The herring-boat sets sail for the grounds in the morning, and at sundown should be at the scene of action. Her nets, aggregating, perhaps, a mile in length, are then "shot," and the boat drifts along towing the line behind her. If fish appear, the nets are hauled in soon after daybreak by the aid of a capstan. The labour of bringing a mile of nets aboard is very severe—so severe, in fact, that the larger boats in many cases employ the help of a small steam-engine. During the return voyage the fish is freed from the meshes, and thrown into the hold ready for sale as soon as land is reached.

Fish, whether for salting or immediate consumption, should be fresh. No class of human food seems to deteriorate so quickly when life is extinct as the "denizens of the deep," so that it is of primary importance to fishermen that their homeward journey should be performed in the shortest possible time. If winds are contrary or absent there may be such delay as to need the liberal use of salt, and even that useful commodity will not stave off a fall in value.

It therefore often happens that a really fine catch arrives at its market in a condition which spells heavy loss to the catchers. A slow return also means missing a day's fishing, which may represent £200 to £300. For this reason the Dogger Bank fishing fleet is served by steam tenders, which carry off the catches as they are made, and thus obviate the necessity for a boat's return to port when its hold is full. Such a system will not, however, be profitable to boats owned by individuals, and working within a comparatively short distance of land.

Each boat must depend on its particular powers, the first to return getting rather better prices than those which come "with the crowd." So steam power is in some cases installed as an auxiliary to the sails, though it may entail the outlay of £2,000 as first cost, and a big bill for upkeep and management. "Small" men cannot afford this expense, and they would be doomed to watch their richer brethren slip into the market before them had not the explosion motor come to their aid. This just meets their case; it is not nearly so expensive to install as steam, occupies much less room, is easier to handle, and therefore saves the expense of trained attendants.

Fishermen are notoriously conservative. To them a change from methods sanctioned by many years of practice is abhorrent. What sufficed for their fathers, they say, should suffice for them. Their trade is so uncertain that a bad season would see no return for the cost of the motor, since, where no fish are caught, it makes little difference whether the journey to port be quick or slow.

However, the motor is bound to come. It has been applied to fishing boats with marked success. While the nets are out, the motor is stopped, and costs not a penny more till the time comes for hauling in. Then it is geared up with a capstan, and saves the crew much of their hardest work. When all is aboard, the capstan hands over the power to the screw, which, together with the sails, propels the vessel homewards at a smart pace. The skipper is certain of making land in good time for the market; and he will be ready for the out voyage next morning. Another point in favour of the motor is that, when storms blow up, the fleet will be able to run for shelter even if the wind be adverse; and we should hear less of the sacrifice of life which makes sad reading after every severe gale.

As to the machinery to be employed, Mr. F. Miller, of Oulton Broad, who first applied the gas-motor to a fishing smack—the *Pioneer*—considers that a 12 h.p. engine would suffice as an auxiliary for small craft of the class found in the northern parts of Great Britain. The Norfolk boats would require a 30 h.p.; and a full-powered boat—*i.e.* one that could depend on the motor entirely—should carry a three-cylinder engine of 80 h.p. In any case, the machinery must be enclosed and well protected; while the lubrication arrangements should be such as to be understood easily by unskilled persons, and absolutely reliable. Owing to the moisture in the atmosphere the ordinary high-tension coil ignition, such as is used on most motor-cars, would not prove efficient, and it is therefore replaced by a low-tension type which makes and breaks the primary circuit by means of a rocking arm working through the walls of the cylinder. Lastly, all parts which require occasional examination or adjustment must be easily accessible, so that they may receive proper attention at sea, and not send the vessel home a "lame duck" under sail.

The advantages of the motor are so great that the Scotch authorities have taken the matter up seriously, appointing an expert to make inquiries. It is therefore quite possible that before many years have elapsed the motor will play an important part in the task of supplying our breakfast tables with the dainty sole or toothsome herring.

A MOTOR FIRE FLOAT

As a good instance of this particular adaptation of the explosion engine to fire-extinction work, we may quote the apparatus now in attendance on the huge factory of Messrs. Huntley and Palmer, the famous Reading biscuit makers. The factory lies along the banks of the river Kennet, which are joined by bridges so close to the water that a steamer could not pass under them. Messrs. Merryweather accordingly built the motor float, 32 feet long, $9\frac{1}{2}$ feet beam, and drawing 27 inches. Two engines, each having four cylinders of a total of 30 h.p., drive two sets of three-cylinder "Hatfield" pumps, which give a continuous feed to the hose. Engines and pumps are mounted on a single bed-plate, and are worked separately, unless it be found advisable to "Siamese" the hoses to feed a single $1\frac{1}{2}$-inch jet, which can be flung to a great height.

One of the most interesting features of the float is the method of propulsion. As its movements are limited to a few hundred yards, the fitting of a screw was considered unnecessary, its place being taken by four jets, two at each end, through which water is forced against the outside water by the extinguishing pumps. These will move the float either forward or astern, steer her, or turn her round.

So here once again petrol has trodden upon the toes of Giant Steam: and very effectively, too.

THE MECHANISM OF THE MOTOR BOAT

In many points the marine motor reproduces the machinery built into cars. The valve arrangements, governors, design of cylinders and water-jackets are practically the same. Small boats carry one cylinder or perhaps two, just as a small car is content with the same number; but a racing or heavy boat employs four, six, and, in one case at least, twelve cylinders, which abolish all "dead points" and enable the screw to work very slowly without engine vibration, as the drive is continuous.

The large marine motor is designed to run at a slower rate than the land motor, and its cylinders are, therefore, of greater size. Some of the cylinders exhibited in the Automobile Show at the London "Olympia" seemed enormous when compared with those doing duty on even high-powered cars; being more suggestive of the parts of an electric lighting plant than of a machine which has to be tucked away in a boat.

Except for the reversing gear, gearing is generally absent on the motor boat. The chauffeur has not to keep changing his speed lever from one notch to another according to the nature of the country. On the sea conditions are more consistently favourable or unfavourable, and, as in a steamboat, speed is controlled by opening or closing the throttle. The screw will always be turned by the machinery, but its effect on the boat must depend on its size and the forces acting in opposition to it. Since water is yielding, it does not offer a parallel to the road. Should a car meet a hill too steep for its climbing powers, the engines must come to rest. The wheel does not slip on the road, and so long as there is sufficient power it will force the car up the severest incline; as soon as the power proves too small for the task in hand the car "lies down." In a motor boat, however, the engine may keep the screw moving without doing more against wind and tide than prevent the boat from "advancing backwards." The only way to make the boat efficient to meet all possible conditions would be to increase the size or alter the pitch of the screw, and to install more powerful engines. "Gearing down"—as in a motor-car—being useless, the only mechanism needed on a motor boat in connection with the transmission of power from cylinders to screw is the reversing gear.

Though engines have been designed with devices for reversing by means of the cams operating the valves, the reversal of the screw's movement is generally effected through gears on the transmission apparatus. The simplest arrangement, though not the most perfect mechanically, is a reversible screw, the blades of which can be made to feather this way or that by the movement of a lever. Sometimes two screws are employed, with opposite twists, the one doing duty while the other revolves idly. But for fast and heavy boats a single solid screw with immovable blades is undoubtedly preferable; its reversal being effected by means of friction clutches. The inelasticity of the explosion motor renders it necessary that the change be made gradually, or the kick of the screw against the motor might cause breakages. The clutch, gradually engaging with a disc revolved by the propeller shaft, first stops the antagonistic motion, and then converts it into similar motion. Many devices have been invented to bring this about, but as a description of them would not be interesting, we pass on to a consideration of the fuel used in the motor boat.

Petrol has the upper hand at present, yet heavier oil must eventually prevail, on account both of its cheapness and of its greater safety. The only objection to its use is the difficulty attending the starting of the engine with kerosene; and this is met by using petrol till the engine and carburetter are hot, and then switching on the petroleum. When once the carburetter has been warmed by exhaust gases to about 270° Fahrenheit it will work as well with the heavy as with the light fuel.

Since any oil or spirit may leak from its tanks and cause danger, an effort has been made to substitute solid for liquid fuel. The substance selected is naphthalene—well known as a protector of clothes against moths. At the "Olympia" Automobile Exhibition of 1905 the writer saw an engine—the Chenier Leon—which had been run with balls of this chemical, fed to the carburetter through a melting-pot. For a description of this engine we must once again have recourse to the *Motor Boat*. The inventors had decided to test its performance with petrol, paraffin, and naphthalene respectively. "The motor, screwed to a testing bench, was connected by the usual belt to a dynamo, so that the power developed under each variety of fuel might be electrically measured, and was then started up on petrol. As soon as the parts were sufficiently warmed up by the exhaust heat, the petrol was turned off, and the motor run for some time on paraffin, until sufficient naphthalene was thoroughly melted to the consistency of a thick syrup. The naphthalene was then fed to its mixing valve through a small pipe dipping into the bottom of the melting-pot, and thence sprayed into the induction chamber to carburate the air therein. Hitherto, the motor had given an average of 12 electrical h.p. at 1,000 revolutions per minute, and it was noticed that as soon as the change was made, this was fully maintained. This test, when continued, bore out others which had previously been made by the firm, and showed the consumption of each of the three fuels to be a little over 12 lbs. per hour for the 12 electrical h.p. given by the motor. Still, the paraffin and naphthalene worked out about equal as to cost, and considering that the latter was in its purest form, as sold for a clothes preservative, we have yet to see how much better its commercial showing will be with lower grades, assuming beforehand that its thermal efficiency and behaviour are as good.

"On the ground of convenience naphthalene, as a solid, is a very long way in front of its liquid rival, kerosene. Its exhaust, too, was much freer from odour, and it appears that, unlike paraffin, it forms neither tar, soot, nor sticky matter, but, on the contrary, has a tendency to brighten all valves, cylinders, walls, etc., any little deposit being a light powder which would be carried into the exhaust."

THE TWO-STROKE MOTOR

In the ordinary "Otto-cycle" motor an explosion occurs once in every two revolutions of the crank. With a single cylinder the energy of the explosion must be stored up in a heavy fly-wheel to carry the engine through the three other operations of scavenging, sucking in a fresh charge, and compressing it preparatory to the next explosion. With two cylinders the fly-wheel can be made lighter, as an explosion occurs every revolution; and in a four-cylinder engine we might almost dispense with the wheel altogether, since the drive is continuous, just as in a double-cylindered steam-engine.

The two-stroke motor, *i.e.* one which makes an explosion for every revolution, is an attempt to unite the advantages of a two-cylindered engine of the Otto type with the lightness of a single-cylindered engine. As it has been largely used for motor boats, especially in America, a short description of its working may be given here.

In the first place, all moving cylinder valves are done away with, their functions being performed by openings covered and opened by the movements of the piston. The crank chamber is quite gas-tight, and has in it a non-return valve through which vapour is drawn from the carburetter every time the piston moves away from the centre. There is also a pipe connecting it with the lower part of the cylinder, but the other end of this is covered by the piston until it has all but finished its stroke.

Let us suppose that an explosion has just taken place. The piston rushes downwards, compressing the gas in the crank chamber to some extent. When the stroke is three-parts performed a second hole, on the opposite side of the cylinder from the aperture already referred to, is uncovered by the piston, and the exploded gases partly escape. Immediately afterwards the second hole is uncovered also, and the fresh charge rushes in from the crank case, being deflected upwards by a plate on the top of the piston, so as to help drive out the exhaust products. The returning piston covers both holes and compresses the charge till the moment of explosion, when the process is repeated. It may be said in favour of this type of engine that it is very simple and free from vibration; against it that, owing to the imperfect scavenging of exploded charges, it does not develop so much power as an Otto-cycle engine of equal cylinder dimensions; also that it is apt to overheat, while it uses double the amount of electric current.

MOTOR BOATS FOR THE NAVY

A country which, like England, depends on the command of the sea for its very existence may well keep a sharp eye on any invention that tends to render that command more certain. In recent years we have heard a lot said, and read a lot written, about the importance of swift boats which in war time could be launched against a hostile fleet, armed with the deadly torpedo. The Russo-Japanese War has given us a fine example of what can be accomplished by daring men and swift torpedo craft.

For some reason or other the British Navy has not kept abreast of France in the number of her torpedo vessels. Reference to official figures shows that, while our neighbours can boast 280 "hornets," we have to our credit only 225. In the House of Commons, on August 10th, 1904, Mr. Henry Norman, M.P., asked the Secretary of the Admiralty whether, in view of the proofs recently afforded of trustworthiness, speed, simplicity, and comparatively low cost of small vessels propelled by petrol motors, he would consider the advisability of testing this class of vessel in His Majesty's Navy. The Secretary replied that the Admiralty had kept a watch on the recent trials and meant to make practical tests with motor pinnaces. In view of the danger that would accompany the storage of petrol on board ship, the paraffin motor was preferable for naval purposes; and an 80 h.p. four-cylindered motor of this type has been ordered from Messrs. Vosper, of Portsmouth.

Mr. Norman, writing in *The World's Work* on the subject, says: "There can be no question that such high speed and cheap construction (80 h.p. giving in the little boat as much speed—to consider that only—as eight thousand in the big boat) point to the use of motor boats for naval purposes in the near future. A torpedo boat exists only to carry one or two torpedoes within launching distance of the enemy. The smaller and cheaper she can be, and the fewer men she carries, provided always she be able to face a fairly rough sea, the better. Now the ordinary steam torpedo boat carries perhaps twenty men, and costs anything from £50,000 to £100,000. A motor boat of equal or greater speed could probably be built for £15,000, and would carry a crew of two men. Six motor boats, therefore, could be built for the cost of one steamboat, and their total crews would not number so many as the crew of the one. Moreover, they could all be slung on board a single vessel, and only set afloat near the scene of action. A prophetic friend of mine declares that the most dangerous warship of the future will be a big vessel, unarmoured and only lightly armed, but of the utmost possible speed, carrying twenty or more motor torpedo boats slung on davits. She will rely on her greater speed for her own safety, if attacked; she will approach as near the scene of action as possible, and will drop all her little boats into the water, and they will make a simultaneous attack. Their hulls would be clean, their machinery in perfect order, their crews fresh and full of energy, and it would be strange if one of the twenty did not strike home. And the destruction of a battleship or great cruiser at the cost of a score of these little wasps, manned by two-score men, would be a very fine naval bargain."

Mr. Norman omits one recommendation that must in active service count heavily in favour of the motor boat, and that is its practical invisibility in the day or at night time. The destroyer, when travelling at high speed, betrays its presence by clouds of smoke or red-hot funnels. The motor boat is entirely free from such dangerous accompaniments; the exhaust from the cylinders is invisible in every way. The very absence of funnels must also be in itself a great advantage. The eye, roving over the waters, might easily "pick up" a series of stumpy, black objects of hard outline; but the motor boat, riding low and flatly on the waves, would probably escape notice, especially when a search-light alone can detect its approach.

It may reasonably be said that the Admiralty knows its own

business best, and that the outsider's opinion is not wanted. The "man in the street" has become notorious for his paper generalship and strategy, and fallen somewhat into disrepute as an adviser on military and naval matters. Yet we must not forget this: that many—we might say most—of the advances in naval mechanisms, armour, and weapons of defence have not been evolved by naval men, but by the highly educated and ingenious civilian who, unblinded by precedent or professional conservatism, can watch the game even better in some respects than the players themselves, and see what the next move should be. That move may be rather unorthodox—like the application of steam to men-o'-war—but none the less the correct one under the circumstances. We allowed other nations to lead us in the matter of breech-loading cannon, armour-plate, submarines, the abolition of combustible material on warships. Shall we also allow them to get ahead with motor boats, and begin to consider that there *may* be something in motor auxiliaries for the fleet when they are already well supplied? If there is a country which should above all others lose no time in adding the motor to her means of defence, that country is Great Britain.

FOOTNOTE:

12. *The Motor Boat*, March 16th, 1905.

CHAPTER IX

THE MOTOR CYCLE

In 1884 the Count de Dion, working in partnership with Messrs. Bouton and Trépardoux, produced a practical steam tricycle. Two years later appeared a somewhat similar vehicle by the same makers which attained the remarkable speed of forty miles an hour. Mr. Serpollet, now famous for his steam cars, built at about the same time a three-wheeled steam tricycle, which also proved successful. But the continuous stoking of the miniature boilers, and the difficulty of keeping them properly supplied with water, prevented the steam-driven cycle from becoming popular; and when the petrol motor had proved its value on heavy vehicles, inventors soon saw that the explosion engine was very much better suited for a light automobile than had been the cumbrous fittings inseparable from the employment of steam.

By 1895 a neat petrol tricycle was on the market; and after the de Dion machines had given proof in races of their capabilities, they at once sprang into popular favour. For the next five years the motor tricycle was a common sight in France, where the excellent roads and the freedom from the restrictions prevailing on the other side of the Channel recommended it to cyclists who wished for a more speedy method of locomotion than unaided legs could give, yet could not afford to purchase a car.

The motor bicycle soon appeared in the field. The earlier types of the two-wheeled motor were naturally clumsy and inefficient. The need of a lamp constantly burning to ignite the charges in the cylinder proved a much greater nuisance on the bicycle than on the tricycle, which carried its driving gear behind the saddle. The writer well remembers trying an early pattern of the Werner motor bicycle in the Champs Elysées in 1897, and his alarm when the owner, while starting the blowlamp on the steering pillar, was suddenly enveloped in flames, which played havoc with his hair, and might easily have caused more serious injuries. Riders were naturally nervous at carrying a flame near the handle-bars, so close to a tank of inflammable petrol liable to leak and catch fire.

The advent of electrical ignition for the gaseous charges opened the way for great improvements, and the motor bicycle slowly but surely ousted its heavier three-wheeled rival. Designs were altered; the engine was placed in or below the frame instead of over the front wheel, and made to drive the back wheel by means of a leather belt. In the earliest types the motive force had either been transmitted by belt to the front wheel, or directly to the rear wheel by the piston rods working cranks on its spindle.

The progress of the motor bicycle has, since 1900, been rapid, and many thousands of machines are now in use. The fact that the engines must necessarily be very small compels all possible saving in weight, and an ability to run continuously at very high speeds without showing serious wear and tear. Details have therefore been perfected, and though at the present day no motor cyclist of wide experience can claim immunity from trouble with his speedy little mount, a really well-designed and well-built machine proves wonderfully efficient, and opens possibilities of locomotion to "the man of moderate means" which were beyond the reach of the rider of a pedal-driven bicycle.

In its way the motor cycle may claim to be one of the most marvellous products of human mechanical skill. Weight has been reduced until a power equal to that of three horses can be harnessed to a vehicle which, when stored with sufficient petrol and electricity to carry it and rider 150 miles, scales about a hundredweight. It will pursue its even course up and down hill at an average of twenty or more miles an hour, the only attention it requires being an occasional charge of oil squirted into the airtight case in which the crank and fly-wheels revolve. The consumption of fuel is ridiculously small, since an economical engine will cover fifteen miles on a pint of spirit, which costs about three-halfpence.

Practically all motor-cycle engines work on the "Otto-cycle" principle. Motors which give an impulse every revolution by compressing the charge in the crank-case or in a separate cylinder, so that it may enter the working cylinder under pressure, have been tried, but hitherto with but moderate success. There is, however, a growing tendency to compass an explosion every revolution by fitting two cylinders, and from time to time four-cylindered cycles have appeared. The disadvantages attending the care and adjustment of so many moving parts has been the cause of four-cylindered cycle motors being unsuccessful from a commercial standpoint, though riders who are prepared to risk extra trouble and expense may find compensation in the quiet, vibrationless drive of a motor which gives two impulses for every turn of the fly-wheel.

The acme of lightness in proportion to power developed has been attained by the "Barry" engine, in which the cylinders and their attachments are made to revolve about a fixed crank, and perform themselves the function of a fly-wheel. So great is the saving of weight that the makers claim a horse-power for every four pounds scaled by the engines; thus, a $3 \frac{1}{2}$ h.p. motor would only just tip the beam against one stone. As the writer has personally inspected a Barry engine, he is able to give a brief account of its action.

It has two cylinders, arranged to face one another on opposite sides of a central air-tight crank-case, the inner end of each cylinder opening into the case. Both pistons advance towards, and recede from, the centre of the case simultaneously. The air-and-gas mixture is admitted into the crank-case through a hole in the fixed crank-spindle, communicating with a pipe leading from the carburetter. The inlet is controlled by a valve, which opens while the pistons are parting, and closes when they approach one another.

We will suppose that the engine is just starting. The pistons are in a position nearest to the crank-case. As they separate they draw a charge—equal in volume to double the cubical contents of one cylinder—into the crank-case through its inlet valve. During the return stroke the charge is squeezed, and passes through a valve into a chamber which forms, as it were, the fourth spoke of a four-spoked wheel, of which the other three spokes are the cylinders and the "silencer." This chamber is connected by pipes to the inlet valves of the cylinders, which are mechanically opened alternately by the action of special cams on the crank-shaft. The cylinder which gets the contents of the compression chamber receives considerably more "mixture" than would flow in under natural suction, and the compression is therefore greater than in the ordinary type of cycle motor, and the explosion more violent. Hence it comes about that the cylinders, which have a bore of only 2 in. and a 2-in. stroke for the piston, develop nearly 2 h.p. each.

It may at first appear rather mysterious how, if the cranks are rigidly attached to the cycle frame, any motion can be imparted to the driving-wheel. The explanation is simple enough: a belt pulley is affixed to one side of the crank-case, and revolves with the cylinders, the silencer, and compression chamber. The rotation is caused by the effort of the piston to get as far as possible away from the closed end of the cylinder after an explosion. Where a crank is movable but the cylinder fixed, the former would be turned round; where the crank is immovable but the cylinder movable, the travel of the piston is possible only if the cylinder moves round the crank. A series of explosions following one another in rapid succession gives the moving parts of the Barry engine sufficient momentum to suck in charges, compress them, and eject the burnt gases. The plan is ingenious, and as the machine into which this type of engine is built weighs altogether only about 70 lbs., the "sport" of motor cycling is open to those people whose age or want of strength would preclude them from the use of the heavy mounts which are still to be seen about the roads. In the future we may expect to find motor cycles approach very closely to a half-hundredweight standard without sacrificing the rigidity needful for fast locomotion over second-class roads.

For "pace-making" on racing tracks, motor cycles ranging up to 24 h.p. have been used; but these are essentially "freak" machines of no practical value for ordinary purposes. Even 3-4 h.p. cycles have set up wonderful records, exceeding fifty miles in the hour, a speed equal to that of a good express train. In comparison with the feats of motor-cars, their achievements may not appear very startling; but when we consider the small size and weight, and the simplicity of the mechanisms which propel cycle and rider at nearly a mile a minute, the result seems marvellous enough.

During the last few years the tricycle has again come into favour, but with the arrangement of its wheels altered; two steering-wheels being placed in front, and a single driving-wheel behind. The main advantage of this inversion is that it permits the fixing of a seat in front of the driver, in which a passenger can be comfortably accommodated. The modern "tricar," with its high-powered, doubled-cylindered engines, its change-speed gears, its friction clutch for bringing the engines gradually into action, its forced water circulation for cooling the cylinders, and its spring-hung frame, is in reality more a car than a cycle, and escapes from the former category only on account of the number of its wheels. To the tourist, or to the person who does not find pleasure in solitary riding, the tricar offers many advantages, and, though decidedly more expensive to keep up than a motor bicycle, entails only very modest bills in comparison with those which affect many owners of cars.

The development of the motor cycle has been hastened and fostered by frequent speed and reliability contests, in which the nimble little motor has acquitted itself wonderfully. A hill a mile long, with very steep gradients, has been ascended in considerably less than two minutes by a $3\frac{1}{4}$ h.p. motor. We read of motor cycles travelling from Land's End to John-o'-Groats; from Calcutta to Bombay; from Sydney to Melbourne; from Paris to Rome—all in phenomenal times considering the physical difficulties of the various routes. Such tests prove the endurance of the motor cycle, and pave the way to its use in more profitable employments. Volunteer cycling corps often include a motor or two, which in active service would be most valuable for scouting purposes, especially if powerful enough to tow a light machine-gun. Commercial travellers, fitting a box to the front of a tricar, are able to scour the country quickly and inexpensively in quest of orders for the firms they represent. The police find the motor helpful for patrolling the roads. On the Continent, and especially in Germany, town and country postmen collect and deliver parcels and letters with the aid of the petrol-driven tricycle, and thereby save much time, while improving the service. Before long, "Hark 'tis the twanging horn" will once again herald the postman's approach in a thousand rural districts, but the horn will not hang from the belt of a horseman, such as the poet Cowper describes, but will be secured to the handle-bars of a neat tricar.

Thus history repeats itself.

Photo] [*Cribb, Southsea.*
A MOTOR LAWN-MOWER

A machine of this kind will cut several acres a day, and also acts as an efficient roller. The operator is able to empty the contents of the catch-box without leaving his seat.

That the motor cycle is still far from perfect almost goes without saying; but every year sees a decided advance in its design and efficiency. The messy, troublesome accumulator will eventually give way to a neat little dynamo, which is driven by the engine and creates current for exploding the cylinder charges as the machine travels. When the cycle is at rest there would then be no fear of electricity leaking away through some secret "short circuit," since the current ceases with the need for it, but starts again when its presence is required. The proper cooling of the cylinders has been made an easier matter than formerly by the introduction of fans which direct a stream of cold air on to the cylinder head. Professor H. L. Callendar has shown in a series of experiments that a fan, which absorbs only 2 to 3 per cent. of an engine's power, will increase the engine's efficiency immensely when a low gear is being used for hill climbing, and the rate of motion through the air has fallen below that requisite to carry off the surplus heat of the motor. If an engine maintains a good working temperature when it progresses through space two feet for every explosion, it would overheat if the amount of progression were, through the medium of a change-gear attachment, reduced to one foot, a change which would be advisable on a steep hill. The fan then supplies the deficiency by imitating the natural rush of air. As Professor Callendar says: "The most important point for the motor cyclist is to secure the maximum of power with the minimum of weight. With this object, the first essentials are a variable speed gear of wide range, and some efficient method of cooling to prevent overheating at low gears.... It is unscientific to double the weight and power of the machine in order to climb a few hills, when the same result can be secured with a variable gear. It is unnecessary to resort to the weight and complication of water cooling when a light fan will do all that is required."

Thus, with the aid of a fan and a gear which will give at least two speeds, the motor cyclist can, with an engine of 2 h.p., climb almost any hill, even without resorting to the help of the pedals. His motion is therefore practically continuous. To be comfortable, he desires immunity from the vibration which quick movement over any but first-class roads sets up in the machine, especially in its forward parts. Several successful spring forks and pneumatic devices have been invented to combat the vibration bogy; and these, in conjunction with a spring pillar for the saddle, which can itself be made most resilient, relieve the rider almost entirely of the jolting which at the end of a long day's ride is apt to induce a feeling of exhaustion. The motor tricycle, which once had a rather bad name for its rough treatment of the nerves, is also now furnished with springs to all wheels, and approximates to the car in the smoothness of its progression.

Assuming, then, that we have motor vehicles so light as to be very manageable, sufficiently powerful to climb severe gradients, reliable, comfortable to ride, and economical in their consumption of fuel and oil, we are able to foresee that they will modify the conditions of social existence. The ordinary pedal-driven cycle has made it possible for the worker to live much further from his work than formerly. "To-morrow, with a motor bicycle, his home may be fifteen miles away, and those extra miles will make a great difference in rent, and in the health of his family. In fact, it almost promises to reconcile the Garden City ideal with the industrial conditions of to-day, by enabling a man to work in the town, and have his home in the country. This advantage applies, of course, less to London than to other great cities, on account of the seemingly endless miles of streets to be traversed before the country is reached. In most manufacturing centres, however, the motoring workman could get to his cottage home by a journey of a few miles. Even in London, moreover, this disadvantage will be overcome to a large extent in the future, for it is as certain as anything of the kind can be that we must ultimately have special highways, smooth, dustless, reserved for motor traffic, leading out of London in the principal directions.... My own conviction is that motor cycling, the simplest, the quickest, the cheapest independent locomotion that has ever been known, is destined to enjoy enormous development. I believe that within a few years the motor bicycle and tricycle will be sold by hundreds of thousands, and that many of the social and industrial conditions of our time will be greatly and beneficially affected by them."[13]

FOOTNOTE:

13. Henry Norman, Esq., M.P., in *The World's Work*.

CHAPTER X

FIRE ENGINES

A GOOD motto to blazon over the doors of a fire-brigade station would be "He gives help twice who gives help quickly." The spirit of it is certainly shown by the brave men who, as soon as the warning signal comes, spring to the engines and in a few minutes are careering at full speed to the scene of operations.

Speed and smartness have for many years past been associated with our fire brigades. We read how horses are always kept ready to be led to the engines; how their harness is dropped on to them and deft fingers set the buckles right in a twinkling, so that almost before an onlooker has time to realise what is happening the sturdy animals are beating the ground with flying hoofs. And few dwellers in large cities have not heard the cry of the firemen, as it rises from an indistinct murmur into a loud shout, before which the traffic, however dense, melts away to the side of the road and leaves a clear passage for the engines, driven at high speed and yet with such skill that accidents are of rare occurrence. The noise, the gleam of the polished helmets, the efforts of the noble animals, which seem as keen as the men themselves to reach the fire, combine to paint a scene which lingers long in the memory.

But efficient as the "horsed" engine is, it has its limitations. Animal strength and endurance are not an indefinite quantity; while the fireman grudges even the few short moments which are occupied by the inspanning of the team. In many towns, therefore, we find the mechanically propelled fire engine coming into favour. The power for working the pumps is now given a second duty of turning the driving-wheels. A parallel can be found in the steam-engine used for threshing-machines, which once had to be towed by horses, but now travels of itself, dragging machine and other vehicles behind it.

The earlier types of automobile fire engines used the boiler's steam to move them over the road. Liverpool, a very enterprising city as regards the extinction of fire, has for some time past owned a powerful steamer, which can be turned out within a minute of the call, can travel at any speed up to thirty miles an hour, and can pump 500 gallons per minute continuously. Its success has led to the purchase of other motor engines, some fitted with a chemical apparatus, which, by the action of acid on a solution of soda in closed cylinders, is enabled to fling water impregnated with carbonic acid gas on to the fire the moment it arrives within working distance of the conflagration, and gives very valuable "first aid" while the pumping apparatus is being got into order.

Two Motor Fire-engines built by Messrs. Merryweather, London. That on the left is driven by petrol, and in addition to pumping-gear carries a wheeled fire-escape. That on the right is driven by steam. Both types are much faster than horses, being able to travel at a rate of over 20 miles an hour.

As might reasonably be expected, the petrol motor has found a fine field for its energies in connection with fire extinction. Since it occupies comparatively little space, more accommodation can be allowed for the firemen and gear. Furthermore, a petrol engine can be started in a few seconds by a turn of a handle, whereas a steamer is delayed until steam has been generated. Messrs. Merryweather have built a four-cylindered, 30 h.p. petrol fire engine capable of a speed of forty miles an hour. It has two systems of ignition—the magneto (or small dynamo) and the ordinary accumulator and coil—so that electrical breakdowns are not likely to occur. A fast motor of this kind, with a pumping capacity of 300 gallons per minute, is peculiarly suited for large country estates, where it can be made to perform household or farm duties when not required for its primary purpose. Considering the great number of country mansions, historically interesting, and full of artistic treasures, which England boasts, it is a matter for regret that such an engine is not always included among the appliances with which every such property is furnished. How often we read "Old mansion totally destroyed by fire," which usually means that in a few short hours priceless pictures, furniture, and other objects of art have been destroyed, because help, when it did come, arrived too late. Owners are, however, more keenly alive to their responsibilities now than formerly. The small hand-worked engine, or the hydrant of moderate pressure, is not considered a sufficient guard for the house and its contents. In many establishments the electric lighting engines are designed to work either the dynamo or a set of pumps as occasion may demand; or the motor is mounted on wheels so that it may be easily dragged by hand to any desired spot.

The "latest thing" in motor fire engines is one which carries a fire-escape with it, in addition to water-flinging machinery. An engine of this type is to be found in some of the London suburbs. A chemical cylinder lies under the driver's seat, where it is well out of the way, and coiled beside it is its reel of hose. The "escape" rests on the top of the vehicle, the wheels hanging over the rear end, while the top projects some distance in front of the steering wheels. The ladder, of telescopic design, can be extended to fifty feet as soon as it has been lowered to the ground. Since the saving of life is even more important than the saving of property, it is very desirable that a means of escape should be at hand at the earliest possible moment after an outbreak. This combination apparatus enables the brigade to nip a fire in the bud, if it is still a comparatively small affair, and also to rescue any people whose exit may have been cut off by the fire having started on or near the staircases.

The Wolseley Motor-Car Company has established a type of chemical motor fire engine which promises to be very successful. A 20 h.p. motor is placed forward under the frame to keep the centre of gravity low. When fully laden, it carries a crew of eight men, two 9-foot ladders, two portable chemical extinguishers, a 50-gallon chemical cylinder, and a reel on which is wound a hose fifty-three yards long. The wheels are a combination of the wooden "artillery" and the wire "spider," wires being strung from the outer end of the hub to the outer ends of the wooden spokes to give them increased power to resist the strain of sudden turns or collisions. An artillery wheel, not thus reinforced, is apt to buckle sideways and snap its spokes when twisted at all.

England has always led the way in matters relating to fire extinction, and to her is due the credit of first harnessing mechanical motive power to the fire engine. Other countries are following her example, and consequently we find fire apparatus moved by the petrol motor in places so far apart as Cape Town, Valparaiso, Mauritius, Sydney, Berlin, New York, Montreal. There can be no doubt but that in a very few years horse-traction will be abandoned by the brigades of our large towns. It has been suggested that the fire-pump of the future will be driven by electricity drawn from switches on the street mains; enough current being stored in accumulators to move the pump from station to fire. In such a case it would be possible to use very powerful pumps, as an electric motor is extremely vigorous for its size and weight. Even to-day steam fire engines can fling 2,000 gallons per minute, and fire floats (for use on the water) considerably more. Possibly the engine of to-morrow will pour 5,000 gallons a minute on the flames if it can get that amount from the water mains, and so render it unnecessary to summon in a large number of engines to quell a big conflagration. Three hundred thousand gallons an hour ought to check a very considerable "blaze."

The force with which a jet of water leaves the huge nozzle of a powerful engine is so great that it would seriously injure a spectator at a distance of fifty yards. The "kick-back" of the water on the nozzle is sometimes sufficient to overcome the power of one man to hold the nozzle in position with his hands, and it becomes needful to provide supports with pointed ends to stick into the ground, or hooks which can be attached to the rungs of a ladder. For an attack on the upper storeys of a house a special "water tower" is much used in America. It consists of a lattice-work iron frame, about twenty-five feet long, inside which slides an extensible iron tube five inches in diameter. The tower is attached to one end of a wagon of unusual length and breadth, and is raised to a vertical position by a rack gearing with a quadrant built into its base below the trunnions or pivots on which it swings. Carbonic acid gas, generated in a cylinder carried on the wagon, works a piston connected with the racks, and on a tap being turned slowly brings the tower to the perpendicular, when it is locked. The telescopic tube, carrying the hose inside it, is then pulled up by windlasses, until the $2\frac{1}{2}$-inch nozzle is nearly fifty feet from the ground. The nozzle itself can be rotated from below by rods and gearing, and the angle of the stream regulated by a rope. If several engines simultaneously deliver their water to the tower hoses 1,000 gallons a minute can be concentrated in a continuous $2\frac{1}{2}$-inch jet on to the fire.

The ordinary horsed fire engine is simple in its design and parts. The vertical boiler contains a number of nearly horizontal water tubes, which offer a great surface to the furnace gases, so that it may raise steam very quickly. The actual water capacity of the boiler is small, and therefore it must be fed continuously by a special pump. The pumps, two or three in number, usually have piston rods working direct from the steam cylinders on the plungers of the pumps. Between cylinders and pumps are slots in the rods in which rotate cranks connected with one another and with a fly-wheel which helps to keep the running steady. After leaving the pumps the water enters a large air vessel, which reduces the sudden shocks of delivery by the cushioning effect of the air, and causes a steady pressure on the water in the hoses.

CHAPTER XI

FIRE-ALARMS AND AUTOMATIC FIRE EXTINGUISHERS

Assuming that a town has a well-appointed fire brigade, equipped with the most up-to-date engines, it still cannot be considered efficiently protected against the ravages of the fire-fiend unless the outbreak of a fire can be notified immediately to the stations, and local mechanical means of suppression come into action almost simultaneously with the commencement of the conflagration. "What you do, do quickly" is the keynote of successful fire-suppression; and its importance has been practically recognised in the invention of hundreds of devices, some of which we will glance at in the following pages.

The electric circuit is the most valuable servant that we have to warn us of danger. Dotted about the streets are posts carrying at the top a circular box, which contains a knob. As soon as a fire is observed, anyone may run to such a post, smash the glass screening the knob, and pull out the latter. This action flashes the alarm to the nearest fire-station, and a few minutes later an engine is dashing to the rescue. Help may also be summoned by means of the ordinary telephone exchanges or from police-stations in direct telephonic communication with the brigade depôts.

All devices depending for their ultimate value on human initiative leave a good deal to be desired. They presuppose conditions which *may* be absent. For instance, an electric wire in a large factory ignites some combustible material during the night. A passer-by may happen to see flames while the fire is in an early stage. On the other hand, it is equally probable that the conflagration may be well established before the alarm is given, with the result that the fire brigade arrives too late to do much good.

What we need, therefore, is a mechanical means of calling attention to the danger automatically, with a quickness which will give the brigade or people close at hand a chance of strangling the monster almost as soon as it is born, and with a precision as to locality that will save the precious time wasted in hunting for the exact point to be attacked.

Mr. G. H. Oatway, M.I.E.E., in a valuable paper read before the International Congress of Fire Brigades in London in 1903, says that the difference between the damage resulting from a fire signalled in its early stage, and the same fire reported when it has spread to two or three floors, is often the difference between a nominal loss and a "burn out." The reformer, he continues, who aims at reducing fire waste must turn his attention primarily to hastening the alarm. The true cure of the matter is, not what quantity of gear it takes to deal with huge conflagrations, but how to concentrate at the earliest stage upon the outbreaks as they occur, and to check them before they have grown beyond control. He cites the fire record of Glasgow of 1902, from which it appears that three fires alone accounted for 40 per cent. of the year's total loss, ten fires for 73 per cent., and the other 706 for only 27 per cent., or an average of £72 per fire. Had the first three fires only been notified at an earlier stage, nearly £72,000 would have been saved. Captain Sir E. M. Shaw, late Chief of the London Fire Brigade, has put the following on record: "Having devoted a very large portion of the active period of my working life in bringing into general use mechanical and hydraulic appliances for dealing with fires after they have been discovered, I nevertheless give and have always given the highest place to the early discovery and indication of fire, and not by any means to the steam, the hydraulic, or the numerous other mechanical appliances on which the principal labours of my life have been bestowed."

A fire given fifteen minutes' start is often hard to overtake. Imagine a warehouse alight on three floors before the alarm is raised! Engines may come one after another and pour deluges of water on the flames, yet as likely as not we read next morning of "total destruction." No stitch in time has saved nine!

The sad part about fires is that they represent so much absolute waste. In commercial transactions, if one party loses the other gains; wealth is merely transferred, and still remains in the community. But in the matter of fire this is not the case. Supposing that a huge cotton mill is burnt down. The re-erection will, it is true, cause a lot of money to change hands; but what has resulted from the money that has *already* been put into the mill? Nothing. So many hundred thousands of pounds have been dematerialised and left nothing behind to represent them. The great Ottawa fire of a few years ago may be remembered as a terrible example of such total loss of human effort.

THE HISTORY OF FIRE-ALARMS

The first recorded specification for an automatic detecting device bears the date 1763. In that year a Mr. John Greene patented an arrangement of cords, weights, and pulleys, which, when the cord burnt through, caused the movement of an indicating semaphore arm. As this action appealed only to the eye, it might easily pass unnoticed, and we can imagine that Mr. Greene did not find a gold mine in his invention.

Twenty-four years later an advance was made when William Stedman introduced a "philosophical fire alarum." "His apparatus consisted of a pivoted bulb having an open neck, and containing mercury, spirit or other liquid. As the heat of the room increased, the expansion of the fluid caused it to spill over, release a trigger, and allow a mechanical gong to run down. This arrangement, whilst an advance upon the first referred to, is quite impracticable. Evaporation of fluid, expansion of mercury, a stiff crank, or other causes which will readily occur to you, and the thing is useless."[14]

In 1806 an automatic method for sprinkling water over a fire appeared. The idea was simplicity itself: a network of water mains, with taps controlled by cords, which burnt through and turned on the water. William Congreve patented, three years later, a sprinkler which was an improvement, in that it indicated the position of the fire in a building by dropping one of a number of weights. But string is not to be relied upon. It may "perish" and break when no fire is about, and any system of extinction depending on it might prove a double-edged weapon.

The nineteenth century produced hundreds of devices for alarming and extinguishing automatically. All depended upon the principle of the expansion or melting of metal in the increased temperature arising from a fire. At one time the circuit-closing thermometer was popular on account of its simplicity. "Its drawback," says Mr. Oatway, "is the smallness of its heat-collecting surface, its isolation, and, last and worst of all, its fixity of operation. In thermometer or fuse-alarm practice it is usual to place the detectors at intervals of about ten feet or so, so that a room of any size will contain a number. If a fire breaks out, the ceiling is blanketed with heat, and every detector feels its influence. Each is affected, but none can give the alarm until some one of the number absolutely reaches the set point or melts out. Having no means of varying the composition of the solder or shifting the wire, an actuating point must be selected which is high enough to give a good working margin over the maximum industrial or seasonal heat of the year; and thus it comes about that if the fire breaks out in winter, or when the room is at its lowest temperature, the amount of loss is considerably and quite unnecessarily increased. In a device set to fuse at 150° Fahrenheit, it will be clear to every one that the measure of the damage will depend upon the normal temperature of the room at the time of the outbreak. If the mercury is in the nineties, there is only some sixty degrees of a rise to wait for; whilst if it happens to be a winter's night, the alarm is held back for a rise of perhaps 120°. What chance is there in this case for a good stop?"

Mr. Oatway has examined the fuses under different conditions, and his conclusions are drawn from practical tests. Great intelligence will not be required to appreciate the force of his arguments. Inasmuch as the rise of temperature caused by a fire is relative, during the early stages at least, to the general heat of the atmosphere, it becomes obvious that an automatic fire-alarm should be one which will keep parallel, as it were, with fluctuations of natural heat. Thus, if the "danger rise" be fixed at 100°, the alarm should be given on a cold night as certainly as at midday in summer. It was the failure of early patterns in this respect that led to their being discredited by the fire-brigade authorities.

The writer already quoted has laid down the functions of a perfect alarm:—

(*a*) To detect the fire at a uniformly early period, under all atmospheric and industrial conditions.

(*b*) To give the alarm upon the premises, and simultaneously to the brigade, by a definite and unmistakable message.

(*c*) To facilitate the work of extinction by indicating the position of the outbreak in the building attacked.

The "May Oatway" alarm has got round the first difficulty in a most ingenious manner by adapting the principle of the compensation methods already described in connection with watches.

The alarm consists of a steel rod of a section found to be most suitable for the purpose. To the side is attached by screws entering the rod near the ends a copper wire, which is long enough to sag slightly at its centre, from which depends a silver chain carrying a carbon contact-piece. A short distance below the carbon are the two terminals of the electric circuit which, when completed by the lowering of the carbon, gives the alarm. Now if there be a very gradual change of temperature the steel rod lengthens slowly, and so does the copper wire, so that the amount of sag remains practically what it was before. But in event of a fire the copper expands much more quickly than the steel, and sags until the carbon completes the circuit. The whole thing is beautifully simple, very durable, quite consistent, and reliable. As soon as the temperature diminishes, on the extinction of the fire, the alarm automatically returns to its normal position, ready for further work.

Now for the second function, that of giving the alarm in many places at once. The closed circuit does not itself directly cause bells to ring: it works a "relay," that is, a second and more powerful circuit. In fact, it is the counterpart of the engine driver, who does not himself make the locomotive move, but merely turns on the steam. An installation has been introduced in the Poplar Workhouse—to quote an instance. Were a fire to break out, one of the 276 detectors would soon set twenty-five bells in action, one in each officer's room. Similarly, in the Warehousemen's Orphanage at Cheadle Hulme, every dormitory would be aroused, and every officer, including the Principal in his house some distance away. Messrs. Arthur and Company, of Glasgow, have a warehouse fortified with 600 of these "nerve centres," all yoked to four position indicators, three of which actuate a "master" indicator connected with the central fire-station. There is no hole or corner in this huge establishment where the fire-demon could essay his fell work without being at once spied upon by a detector.

We may glance for a moment at the mechanism which sends an unmistakable message for help. At the brigade station there is a number of small tablets, each protected by a flap, on the outside of which is the word **SAFE**, on the inside **FIRE**. Normally the flap is closed. As soon as the circuit is completed, a magnet releases the flap, and a bell begins to ring. Now, it is possible that the circuit might be closed accidentally by contact somewhere between the premises it serves and the fire-station. So that the official on guard, seeing "J. Brown and Company" on the uncovered tablet, might despatch the engines to the place indicated on a wild-goose chase.

To prevent such false alarms the transmitter not only rings the station up, but automatically sends an unmistakable message. When a fire occurs an automatic printing machine is set in motion to despatch a cipher in the Morse code *four times* to the station. An accidental circuit could not do this; therefore, when the officer sees on the receiving tape the well-known cipher, he turns out his men with all speed.

On arriving at their destination the firemen receive valuable help from the "position indicator," which guides them to their work. On a special board is seen a row, or rows, of shutters similar to those already mentioned. Each row belongs to a floor; each unit of the row to a room. A glance suffices to tell that the trouble is, say, in the most southerly room of the second floor. No notice is therefore taken of smoke rolling out of other parts of the building, until the danger spot has been attacked.

That the firemen appreciate such an ally goes without saying. Every fire extinguished is a point to their credit. Also, the risks they run are greatly diminished, while the wear and tear of tackle is proportionately reduced. The fireman is noted for his courage and unflinching performance of duty. The discomforts of his profession are sometimes severe, and its dangers as certain as they are at times appalling. Therefore we welcome any mechanical method which at once shortens his work, lessens his peril, and protects property from damage.

Mr. Oatway draws special attention to the need for simultaneous warning on the premises and at the fire-station. "I remember," he says, "many cases, but perhaps no better illustration need be looked for than the case of a cotton mill in Lancashire about two years ago (1901). The fire was seen to start at a few minutes past seven; a fuse blew out, and sparked some cotton; but it looked such a simple job that the operatives elected to deal with it. At twenty minutes to eight it dawned upon somebody that the brigade had better be sent for, because the fire was getting away; and in due course they arrived; but the mill, already doomed, became a total loss. In every centre similar instances can be quoted. There is nothing in any automatic system to discourage individual effort. Inmates can put the fire out, if able; but in any case the brigade gets timely and definite notice, and if on their arrival they find the fire extinguished, as Chief Superintendent Thomas put it when we opened the Dingle Station after the fatal train-burning, 'So much the better, we shall get to our beds all the quicker.' This is the common-sense view of it. Helpers work none the less intelligently because they know the brigade is coming; and it is necessary to provide some automatic method of calling them, because you can never rely upon anybody who is unfamiliar with fire doing the right thing at the proper time."

Messrs. May and Oatway, who give their name to the alarm described above, first introduced their apparatus in New Zealand, from which country it has spread over the British Empire. The largest installation is at Messrs. Clark and Company's Anchor Mills, Paisley. The whole of the immense block of buildings, the greater part of which was previously protected by "sprinklers" only, is now electrically protected also; and connected up with the fire brigade, and through their station with the sleeping quarters of every fireman. Some figures will be interesting here. There are 119 *miles* of internal alarm circuits; $5\frac{1}{4}$ miles of underground cable between buildings; 19 automatic telegraphs;

21 automatic position indicators; 20 alarm gongs a foot in diameter.

Early in January, 1905, a fire broke out in these buildings during the dinner hour, when most of the works' firemen were at their midday meal. The alarm sounded simultaneously at the works' fire-station and at the firemen's houses, which are situated on the other side of the street from the mill. The firemen were on the spot immediately, and were enabled to subdue the flames, which had broken out in the building occupied as warehouse and office, before it had got a firm hold of the inflammable material, although not before one of the large stacks of finished thread was ablaze. The brigade, however, were soon masters of the situation, and the damage done was under £100. There is little doubt, had the alarm been left to the ordinary course, the building would have been totally destroyed.[15]

In those few minutes the installation saved its entire cost many times over. Truly

> "A little fire is quickly trodden out,
> Which, being suffered, rivers cannot quench."

Here, in a Shakespearean nutshell, is the whole science of fire protection.

AUTOMATIC SPRINKLERS

As these have been referred to several times a short description may appropriately be given. The building which they protect is fitted with a network of mains and branches ramifying into each room. At the end of each branch is a nozzle, the mouth of which is bridged over by a metal arch carrying a small plate. Between the bridge and a glass plug closing the nozzle is a bar of easily fusible solder. When the temperature has risen to danger point the solder melts, and the plug is driven out by the water, which strikes the plate and scatters in all directions.

This device has proved very valuable on many occasions. The *Encyclopædia Britannica* (Tenth Edition) states that, in the record of the American Associated Factory Mutual companies for the $5\frac{1}{2}$ years ending January 1, 1900, it appears that out of 563 fires where sprinklers came into play 129 were extinguished by one jet; 83 by two jets; 61 by three; 44 by four; 40 by five.

The fire-bucket is the simplest device we have as a first aid; and very effective it often proves. Insurance statistics show that more fires are put out by pails than by all other appliances put together. The important point to be remembered in connection with them is that they should always *be kept full*; so that, at the critical moment, there may be no hurried rushing about to find the two gallons of liquid which each is supposed to contain permanently. In *Cassier's Magazine* (vol. xx. p. 85) is given an account of the manner in which an ingenious mill superintendent ensured the pails on the premises being ready for duty. The hooks carrying the pails were fitted up with pieces of spring steel strong enough to lift the pail when nearly empty, but not sufficiently so to lift a full pail. Just over each spring, in such a position as to be out of the way of the handle of the pail, was set a metal point, connected with a wire from an open-circuit battery. So long as the pails were full, their weight, when hung on their hooks, kept the springs down, but as soon as one was removed, or lost a considerable part of its contents by evaporation or otherwise, the spring on its hook would rise, come into contact with the metal point, thus close the battery circuit and ring a bell in the manager's office, at the same time showing which was the bucket at fault. The bell continued to ring till the deficiency had been made right; and by this simple contrivance the buckets were protected from misuse or lack of attention.

FOOTNOTES:

14. Mr. W. H. Oatway.

15. *Glasgow Evening News.*

CHAPTER XII

THE MACHINERY OF A SHIP

THE REVERSING ENGINE — MARINE ENGINE SPEED GOVERNORS — THE STEERING ENGINE — BLOWING AND VENTILATING APPARATUS — PUMPS — FEED HEATERS — FEED-WATER FILTERS — DISTILLERS — REFRIGERATORS — THE SEARCH-LIGHT — WIRELESS TELEGRAPHY INSTRUMENTS — SAFETY DEVICES — THE TRANSMISSION OF POWER ON A SHIP

WITH many travellers by sea the first impulse, after bunks have been visited and baggage has been safely stored away, is to saunter off to the hatches over the engine-room and peer down into the shining machinery which forms the heart of the vessel. Some engine is sure to be at work to remind them of the great power stored down there below, and to give a foretaste of what to expect when the engine-room gong sounds and the man in charge opens the huge throttle controlling some thousands of horse-power.

By craning forward over the edge of the ship, a jet of water may be seen spurting from a hole in the side just above the water-line, denoting either that a pump is emptying the bilge, or that the condensers are being cooled ready for the work before them.

Towards the forecastle a busy little donkey engine is lifting bunches of luggage off the quay by means of a rope passing over a swinging spar attached to the mast, and lowering it into the nether regions where stevedores pack it neatly away.

In a small compartment on the upper deck is some mysterious, and not very important-looking, gear: yet, as it operates the rudder, it claims a place of honour equalling that of the main engines which turn the screw.

To the ordinary passenger the very existence of much other machinery—the reversing engines, the air-pumps, the condensers, the "feed" heaters, the filters, the evaporators and refrigerators, and the ventilators—is most probably unsuspected. The electric light he would, from his experience of things ashore, vaguely connect with an engine "somewhere." But the apparatus referred to either works so unobtrusively or is so sequestered from the public eye that one might travel for weeks without even hearing mention of it.

On a warship the amount of machinery is vastly increased. In fact, every war vessel, from the first-class battleship to the smallest "destroyer," is practically a congeries of machines; accommodation for human beings taking a very secondary place. Big guns must be trained, fed, and cleaned by machinery; and these processes, simple as they sound, need most elaborate devices. The difference in respect of mechanism between the *King Edward VII.* and Nelson's *Victory* is as great as that between a motor-car and a farmer's cart. It would not be too much to say that the mechanical knowledge of any period is very adequately gauged from its fighting vessels.

Photo] [Cribb, Southsea.
A gigantic sheer-legs used for lowering boilers, big guns, turrets, etc., into men-of-war. The legs rise to a height of 140 feet, and will handle weights up to 150 tons.

During the last twenty years marine engines have been enormously improved. But the advance of auxiliary appliances has been even more marked. In earlier times the matter considered of primary importance was the propulsion of the vessel; and engineers turned their attention to the problem of crowding the greatest possible amount of power into the least possible amount of space. This was effected mainly by the "compounding" of engines—using the steam over and over again in cylinders of increasing size—and by improving the design of boilers. As soon as this business had been well forwarded, auxiliary machinery, which, though not absolutely necessary for movement, greatly affected the ease, comfort, and economy of working a ship, got its share of notice, with the result that a tour round the "works" of a modern battleship or liner is a growing wonder and a liberal education in itself.

This chapter will deal with the auxiliaries to be found in large vessels designed for peaceful or warlike uses. Many devices are common to ships of both classes, and some are confined to one type only, though the "steel wall" certainly has the advantage with regard to multiplicity.

We may begin with

THE REVERSING ENGINE

All marine engines should be fitted with some apparatus which enables the engineer to reverse them from full speed ahead to full speed astern in a few seconds. The effort required to perform the operation of shifting over the valves is such as to necessitate the help of steam. Therefore you will find a special device in the engine-room which, when the engineer moves a small lever either way from the normal position, lets steam into a cylinder and moves rods reversing the main engine. By a link action (which could not be explained without a special diagram) the valves of the auxiliary are closed automatically as soon as the task has been performed; so that there is no constant pressure on the one or the other side of its piston. To prevent the reversal being too sudden, the auxiliary's piston-rod is prolonged, and fitted to a second piston working in a second cylinder full of glycerine or oil. This piston is pierced with a small hole, through which the incompressible liquid passes as the piston moves. Since its passage is gradual, the engines are reversed deliberately enough to protect their valves from any severe strains. These reversing engines can, if the steam serving them fails, be worked by hand.

MARINE ENGINE SPEED GOVERNORS

When a ship is passing through a strong sea and pitches as she crosses the waves, the screw is from time to time lifted clear of the water, and the engines which a moment before had been doing their utmost, suddenly find their load taken off them. The result is "racing" of the machinery, which makes itself very unpleasantly felt from one end of the ship to the other. Then the screw, revolving at a speed much above the normal, suddenly plunges into the water again, and encounters great resistance to its revolution.

A series of changes from full to no "load," as engineers term it, must be harmful to any engines, even though the evil effects are not shown at once. Great strains are set up which shake bolts loose, or may crack the heavy standards in which the cranks and shaft work, and even seriously tax the shaft itself and the screw. On land every stationary engine set to do tasks in which the load varies—which practically means all stationary engines—are fitted with a governor, to cut off the steam directly a certain rate of revolution is exceeded. These engines are the more easily governed because they carry heavy fly-wheels, which pick up or lose their velocity gradually. A marine engine, on the other hand, has only the screw to steady it, and this is extremely light in proportion to the power which drives it; in fact, has scarcely any controlling influence at all as soon as it leaves the water.

Marine engineers, therefore, need some mechanical means of restraining their engines from "running away." The device must be very sensitive and quick acting, since the engines would increase their rate threefold in a second if left ungoverned when running "free"; while on the other hand it must not throttle the steam supply a moment after the work has begun again when the screw takes the water.

Many mechanisms have been invented to curb the marine engine. Some have proved fairly successful, others practically useless; and the fact remains that, owing to the greater difficulty of the task, marine governing is not so delicate as that of land engines. A great number of steamships are not fitted with governors, for the simple reason that the engineers are sceptical about such devices as a class and "would rather not be bothered with them."

But whatever may have been its record in the past, the marine governor is at the present time sufficiently developed to form an item in the engine-rooms of many of our largest ships. We select as one of the best devices yet produced that known as Andrews' Patent Governor; and append a short description.

It consists of two main parts—the pumps and the ram closing the throttle. The pumps, two in number, are worked alternately by some moving part of the engine, such as the air-pump lever. They inject water through a small pipe into a cylinder, the piston-rod of which operates a throttle valve in the main steam supply to the engines. At the bottom of this cylinder is a by-pass, or artificial leak, through which the water flows back to the pumps. The size of the flow through the by-pass is controlled by a screw adjustment.

We will suppose that the governor is set to permit one hundred revolutions a minute. As long as that rate is not exceeded the by-pass will let out as much water as the pumps can inject into the cylinder, and the piston is not moved. But as soon as the engines begin to race, the pumps send in an excess, and the piston immediately begins to rise, closing the throttle. As the speed falls, the leak gets the upper hand again, and the piston is pushed down by a powerful spring, opening the throttle.

It might be supposed that, when the screw "races," the pumps would not only close the throttle, but also press so hard on it as to cause damage to some part of the apparatus before the speed had fallen again. This is prevented by the presence of a second control valve (or leak) worked by a connecting-rod rising along with the piston-rod of the ram. The two rods are held in engagement by a powerful spring which presses them together, so that a hollow in the first engages with a projection on the second. Immediately the pressure increases and the piston rises, the second valve is shut by the lifting of its rod, and so farther augments the pressure in the cylinder and quickens the closing of the throttle valve. This pressure increase must, however, be checked, or the piston would overrun and stop the engines. So when the piston has nearly finished its stroke the connecting-rod comes into contact with a stop which disengages it from the piston-rod and allows the second control valve to be fully opened by the spring pulling on its rod. The piston at once sinks to such a position as the pressure allows, and the action is repeated time after time.

The governing is practically instantaneous, though without shock, and is said to keep the engine within 3 per cent. of the normal rate. That is, if 100 be the proper number of revolutions, it would not be allowed to exceed 103 or drop below 97. Such governing is, in technical language, very "close."

The idea is very ingenious: pumps working against a leak, and as soon as they have mastered it, being aided by a secondary valve which reduces the size of the leak so as to render the effect of the pumps increasingly rapid until the throttle has been closed. Then the secondary valve is suddenly thrown out of action, gives the leak full play, and causes the throttle to open quickly so that the steam may be cut off only for a moment. By the turning of a small milled screw-head a couple of inches in diameter the pace of 5,000 h.p. engines is as fully regulated as if a powerful brake were applied the moment they exceeded "the legal limit."

STEERING ENGINES

The uninitiated may think that the man on the bridge, revolving a spoked-wheel with apparently small exertion, is directly moving the rudder to port or to starboard as he wishes. But the helm of a large vessel, travelling at high speed, could not be so easily deflected were not some giant at work down below in obedience to the easy motions of the wheel.

Sometimes in a special little cabin on deck, but more often in the engine-room, where it can be tended by the staff, there is the steering engine, usually worked by steam-power. Two little cylinders turn a worm-screw which revolves a worm-wheel and a train of cogs, the last of which moves to right or left a quadrant attached to the chains or cables which work the rudder. All that the steersman has to do with his wheel is to put the engine in forward, backward, or middle gear. The steam being admitted to the cylinders quickly moves the helm to the position required.

A particularly ingenious steam gear is that made by Messrs. Harfield and Company, of London. Its chief feature is the arrangement whereby the power to move the rudder into any position remains constant. If you have ever steered a boat, you will remember that, when a sudden curve must be made, you have to put far more strength into the tiller than would suffice for a slight change of direction. Now, if a steam-engine and gear were so built as to give sufficient pressure on the helm in all positions, it would, if powerful enough to put the ship hard-a-port, evidently be overpowered for the gentler movements, and would waste steam. The Harfield gear has the last of the cog-train—the one which engages with the rack operating the tiller—mounted eccentrically. The rack itself is not part of a circle, but almost flat centrally, and sharply bent at the ends. In short, the curve is such that the rack teeth engage with the eccentric cog at all points of the latter's revolution.

When the helm is normal the longest radius of the eccentric is turned towards the rack. In this position it exerts least power; but least power is then needed. As the helm goes over, the radius of the cogs gradually decreases, and its leverage proportionately increases. So that the engine is taxed uniformly all the time.

Some war vessels, including the ill-fated Russian cruiser *Variag*, have been fitted with electric steering gear, operated by a motor in which the direction of the current can be varied at the will of the helmsman.

All power gears are so arranged that, in case of a breakdown of the power, a hand-wheel can be quickly brought into play.

BLOWING AND VENTILATING APPARATUS

A railway locomotive sends the exhaust steam up the funnel with sufficient force to expel all air from the same and to create a vacuum. The only passage for the air flying to fill this empty space lies through the fire-box and tubes traversing the boiler from end to end. Were it not for the "induced draught"—the invention of George Stephenson—no locomotive would be able to draw a train at a higher speed than a few miles an hour.

On shipboard the fresh water used in the boilers is far too precious to be wasted by using it as a fire-exciter. Salt water to make good the loss would soon corrode the boilers and cause terrible explosions. Therefore the necessary draught is created by *forcing* air through the furnaces instead of by *drawing* it.

The stoke-hold is entirely separated from the outer air, except for the ventilators, down which air is forced by centrifugal pumps at considerable pressure. This draught serves two purposes. It lowers the temperature of the stoke-hold, which otherwise would be unbearable, and also feeds the fires with plenty of oxygen. The air forced in can escape in one way only, viz. by passing through the furnaces. When the ship is slowed down the "forced draught" is turned off, and then you see the poor stokers coming up for a breath of fresh air. In the Red Sea or other tropical latitudes these grimy but useful men have a very hard time of it. While passengers up above are grumbling at the heat, the stoker below is almost fainting, although clad in nothing but the thinnest of trousers.

In the engine-room also things at times become uncomfortably warm. Take the case of the United States monitor *Amphitrite*, which went into commission in 1895 for a trial run.

Both stoke-hold and engine-room were very insufficiently ventilated. The vessel started from Hampton Roads for Brunswick, Georgia. "The trip of about 500 miles occupied five days in the latter part of July, and, for sheer suffering, has perhaps seldom been equalled in our naval history. The fire-room (stoke-hold) temperature was never below 150°, and often above 170°, while the engine-room ranged closely about 150°. For the first twenty-four hours the men stood it well, but on the second day seven succumbed to the heat and were put on the sick list, one of them nearly dying; before the voyage was ended, twenty-eight had been driven to seek medical attendance. The gaps thus created were partially filled with inexperienced men from the deck force, until there was only a lifeboat's crew left in each watch.... On the evening of the fourth day out our men had literally fought the fire to a finish and had been vanquished; the watch on duty broke down one by one, and the engines, after lumbering along slower and slower, actually stopped for want of steam.... At daybreak the next morning we got under way and steamed at a very conservative rate to our destination, fortunately only about ten miles distant. The scene in the fire-room that morning was not of this earth, and far beyond description. The heat was almost destructive to life; steam was blowing from many defective joints and water columns; tools, ladders, doors, and all fittings were too hot to touch; and the place was dense with smoke escaping from furnace doors, for there was absolutely no draught. The men collected to build up the fires were the best of those remaining fit for duty, but they were worn out physically, were nervous, apprehensive, and dispirited. Rough Irish firemen, who would stand in a fair fight till killed in their tracks, were crying like children, and begging to be allowed to go on deck, so completely were they unmanned by the cruel ordeal they had endured so long. 'Hell afloat' is a nautical figure of speech often idly used, but then we saw it. For a month thereafter the ship was actively employed on the southern coast, drilling militia at different ports, and sweltering in the new dock at Port Royal. One trip of twenty-nine hours broke the record for heat, the fire-room being frequently above 180°. All fire-room temperatures were taken in the actual spaces where the men had to work, and not from hot corners or overhead pockets."[16]

The ventilators were subsequently altered, and the men enjoyed comparative comfort. The words quoted will suffice to establish the importance of a proper current of air where men have to work. One of the greatest difficulties encountered in deep mining is that, while the temperature approaches and sometimes passes that of a stoke-hold, the task of sending down a cool current from above is, with depths of 4,000 ft. and over, a very awkward one to carry out.

On passenger ships the fans ventilating the cabins and saloons are constantly at work, either sucking out foul air or driving in fresh. The principle of the fan is very similar to that of the centrifugal water pump—vanes rotating in a case open at the centre, through which the air enters, to be flung by the blades against the sides of the case and driven out of an opening in its circumference. Sometimes an ordinary screw-shaped fan, such as we often see in public buildings, is employed.

PUMPS

Every steamship carries several varieties of pump. First, there are the large pumps, generally of a simple type, for emptying the bilge or any compartment of the ship which may have sprung a leak. "All hands to the pumps!" is now seldom heard on a steamer, for the opening of a steam-cock sets machinery in motion which will successfully fight any but a very severe breach. It is needless to say that these pumps form a very important part of a ship's equipment, without which many a fine vessel would have sunk which has struggled to land.

The pumps for the condensers form another class. These are centrifugal force pumps; their duty is to circulate cold sea-water round the nests of tubes through which steam flows after passing through the cylinders. It is thus converted once more into water, ready for use again in the boiler. Every atom of the water is evaporated, condensed, and pumped back into the boiler once in a period ranging from fifteen minutes to an hour, according to the type of boiler and the size of the supply tanks.

Some condensers have the cooling water passed through the tubes, and the steam circulated round these in an air-tight chamber. In any case, the condenser should be so designed as to offer a large amount of cold surface to the hot vapour. A breakdown of the condenser pumps is a serious mishap, since steam would then be wasted, which represents so much fresh water—hard to replace in the open sea. It would be comparable to the disarrangement of the circulating pump on a motor-car, though the effects are different.

We must not forget the feed-pumps for the boilers. On their efficient action depends the safety of the ship and her passengers. Water must be maintained at a certain level in the boiler, so that all tube and other surfaces in direct contact with the furnace gases may be covered. The disastrous explosions we sometimes hear of are often caused by the failure of a pump, the burning of a tube or plate, and the inevitable collapse of the same. The firms of Weir and Worthington are among the best-known makers of the special high-pressure pumps used for throwing large quantities of water into the boilers of mercantile and war vessels.

FEED HEATERS

As the fuel supply of a vessel cannot easily be replenished on the high seas, economy in coal consumption is very desirable.

If you put a cold spoon into a boiling saucepan ebullition is checked at once, though only for a moment, while the spoon takes in the temperature of the water. Similarly, if cold water be fed into a boiler the steam pressure at once falls. Therefore the hotter the feed water is the better.

The feed heater is the reverse of the condenser. In the latter, cold water is used to cool hot steam; in the former, hot steam to heat cold water. There are many patterns of heaters. One type, largely used, sprays the cold water through a valve into a chamber through which steam is passed from the engines. The spray, falling through the hot vapour, partially condenses it and takes up some of its heat. The surplus steam travels on to the condensers. A float in the lower part of the chamber governs a valve admitting steam to the boiler pumps, so that as soon as a certain amount of water has accumulated the pumps are started, and the hot liquid is forced into the boiler.

Another type, the Hampson feeder, sends steam through pipes of a wavy form surrounded by the feed water, there being no actual contact between liquid and vapour.

An ally of the heater is the

FEED-WATER FILTER,

which removes suspended matter which, if it entered the boiler, would form a deposit round the tubes, and while decreasing their efficiency, make them more liable to burning. The most dangerous element caught by the filters is fatty matter—oil which has entered the cylinders and been carried off by the exhaust steam.

The filter is either high pressure, *i.e.* situated between the pump and the boiler; or low pressure, *i.e.* between the pump and the reservoir from which it draws its water. The second class must have large areas, so as not to throttle the supply unduly.

Many kinds of filtering media have been tried—fabrics of silk, calico, cocoanut fibre, towelling, sawdust, cork dust, charcoal, coke; but the ideal substance, at once cheap, easily obtainable, durable, and completely effective, yet remains to be found.

A filter should be so constructed that the filtering substance is very accessible for cleansing or renewal.

DISTILLERS

We now come to a part of a ship's plant very necessary for both machines and human beings. Many a time have people been in the position of the Ancient Mariner, who exclaimed:—

"Water, water, everywhere,
But not a drop to drink!"

Water is so weighty that a ship cannot carry more than a very limited quantity, and that for the immediate needs of her passengers. The boilers, in spite of their condensers, waste a good deal of steam at safety valves through leaking joints and packings, and in other ways. This loss must be made good, for, as already remarked, salt water spells the speedy ruin of any boiler it enters.

The distiller in its simplest form combines a boiler for changing water into vapour, with a condenser for reconverting it to liquid. Solids in impure water do not pass off with the steam, so that the latter, if condensed in clean vessels, is fit for drinking or for use in the engine boilers.

A pound of steam will, under this system, give a pound of water. But as such procedure would be extravagant of fuel, *compound* condensers are used, which act in the following manner.

High-pressure steam is passed from the engine boilers into the tubes of an evaporator, and converts the salt water surrounding it into steam. The boiler steam then travels into its own condenser or into the feed water heater, while the steam it generated passes into the coils of a second evaporator, converts water there into steam, and itself goes to a condenser. The steam generated in the second evaporator docs similar duty in a third evaporator. So that one pound of high-pressure steam is directly reconverted to water, and also indirectly produces between two and three pounds of fresh water.

The condensers used are similar to those already described in connection with the engines, and need no further comment. About the evaporators, it may be said that they are so constructed that they can be cleaned out easily as soon as the accumulation of salt and other matter renders the operation necessary. Usually one side is hinged, and provided with a number of bolts all round the edges which are quickly removed and replaced.

The United States Navy includes a ship, the *Iris*, whose sole duty is to supply the fleet she attends with plenty of fresh water. She was built in 1885 by Messrs. R. and W. Hawthorn, of Newcastle-on-Tyne, and measures 310 feet in length, $38\frac{1}{2}$ feet beam. For her size she has remarkable bunker capacity, and can accommodate nearly 2,500 tons of coal. Fore and aft are huge storage tanks to hold between them about 170,000 gallons of fresh water. Her stills can produce a maximum of 60,000 gallons a day. It has been reckoned that each *ton* of water distilled costs only 18 cents; or, stated otherwise, that 40 gallons cost one penny. At many ports fresh water costs three or four times this figure; and even when procured is of doubtful purity. During the Spanish-American War the *Iris* and a sister ship, the *Rainbow*, proved most useful.

REFRIGERATORS

Of late years the frozen-meat trade has increased by leaps and bounds. Australia, New Zealand, Argentina, Canada, and the United States send millions of pounds' worth of mutton and beef across the water every year to help feed the populations of England and Europe.

In past times the live animals were sent, to be either killed when disembarked or fatted up for the market. This practice was expensive, and attended by much suffering of the unfortunate creatures if bad weather knocked the vessel about.

Refrigerating machinery has altered the traffic most fundamentally. Not only can more meat be sent at lower rates, but the variety is increased; and many other substances than flesh are often found in the cold stores of a ship—butter and fruit being important items.

Certain steamship lines, such as the Shaw, Savill, and Albion—plying between England and Australasia—include vessels specially built for the transport of vast numbers of carcases. Upwards of a million carcases have been packed into the hull of a single ship and kept perfectly fresh during the long six weeks' voyage across the Equator.

Every passenger-carrying steamer is provided with refrigerating rooms for the storage of perishable provisions; and as the comfort of the passengers, not to say their luxury, is bound up with these compartments, it will be interesting to glance at the method employed for creating local frost amid surrounding heat.

The big principle underlying the refrigerator is this—that a liquid when turned into gas *absorbs* heat (thus, to convert water into steam you must feed it with heat from a fire), and that as soon as the gas loses a certain amount of its heat it reverts to liquid form.

Now take ammonia gas. The "spirits of hartshorn" we buy at the chemist's is water impregnated with this gas. At ordinary living temperatures the water gives out the gas, as a sniff at the bottle proves in a most effective manner.

If this gas were cooled to 37·3° below zero it would assume a liquid state, *i.e.* that temperature marks its boiling point. Similarly steam, cooled to 212° Fahr., becomes water. Boiling point, therefore, merely means the temperature at which the change occurs.

Ammonia liquid, when gasifying, absorbs a great amount of heat from its surroundings—air, water, or whatever they may be. So that if we put a tumbler full of the liquid into a basin of water it would rob the water of enough heat to cause the formation of ice.

The refrigerating machine, generally employed on ships, is one which constantly turns the ammonia liquid into gas, and the gas back into liquid. The first process produces the cold used in the freezing-rooms. The apparatus consists of three main parts:—

(1) The *compressor*, for squeezing ammonia gas.

(2) The *condenser*, for liquefying the gas.

(3) The *evaporator*, for gasifying the liquid.

The *compressor* is a pump. The *condenser*, a tube or series of tubes outside which cold water is circulated. The *evaporator*, a spiral tube or tubes passing through a vessel full of brine. Between the condenser and evaporator is a valve, which allows the liquid to pass from the one to the other in proper quantities.

We can now watch the cycle of operations. The compressor sucks in a charge of very cold gas from the evaporator, and squeezes it into a fraction of its original volume, thereby heating it. The heated gas now passes into the condenser coils and, as it expands, encounters the chilling effects of the water circulating outside, which robs it of heat and causes it to liquefy.

It is next slowly admitted through the expansion valve into the evaporator. Here it gradually picks up the heat necessary for its gaseous form: taking it from the brine outside the coils, which has a very low freezing-point. The brine is circulated by pumps through pipes lining the walls of the freezing-room, and robs the air there of its heat until a temperature somewhat below the freezing-point of water is reached.

The room is well protected by layers of charcoal or silicate cotton, which are very bad conductors of heat. How the chamber strikes a novice can be gathered from the following description of a Cunard liner's refrigerating room. "It is a curious and interesting sight. It may be a hot day on deck, nearing New York, and everyone is going about in sun hats and light clothes. We descend a couple of flights of stairs, turn a key, and here is winter, sparkling in glassy frost upon the pale carcases of fowls and game, and ruddy joints of meat, crystallising the yellow apples and black grapes to the likeness of sweetmeats in a grocer's shop, gathering on the wall-pipes in scintillating coats of snow nearly an inch deep. You can make a snowball down here, if you like, and carry it up on deck to astonish the languid loungers sheltering from the sun under the protection of the promenade-deck roof. Such is the modern substitute for the old-time salt-beef cask and bags of dried pease!"

The larder is so near the kitchen that while below decks we may just peep into the kitchens, where a white-capped *chef* presides over an army of assistants. Inside a huge oven are dozens of joints turning round and round by the agency of an invisible electric-motor. But what most tickles the imagination is an electrical egg-boiling apparatus, which ensures the correct amount of cooking to any egg. A row of metal dippers, with perforated bottoms, is suspended over a trough of boiling water. Each dipper is marked for a certain time—one minute, two, three, four, and so on. The dippers, filled with eggs, are pushed down into the water. No need to worry lest they should be "done to a bullet," for at the expiry of a minute up springs the one-minute dipper; and after each succeeding minute the others follow in due rotation. Where 2,000 eggs or more are devoured daily this ingenious automatic device plays no mean part.

THE SEARCH-LIGHT

All liners and war vessels now carry apparatus which will enable them to detect danger at night time, whether rocks or an enemy's fleet, icebergs or a water-logged derelict. On the bridge, or on some other commanding part of the vessel's structure, is a circular, glass-fronted case, backed with a mirror of peculiar shape. Inside are two carbon points almost touching, across which, at the turn of a handle, leaps a shower of sparks so continuous as to form a dazzling light. The rays from the electric arc, as it is called, either pass directly through the glass lens, or are caught by the parabolic reflector and shot back through it in an almost parallel pencil of wonderful intensity, which illumines the darkness like a ray of sunshine slanting through a crack in the shutter of a room. The search-light draws its current from special dynamos, which absorb many horse-power in the case of the powerful apparatus used on warships. At a distance of several miles a page of print may be easily read by the beams of these scrutinisers of the night.

The finest search-lights are to be found ashore at naval ports, where, in case of war, a sharp look-out must be kept for hostile vessels. Portsmouth boasts a light of over a million candle-power, but even this is quite eclipsed by a monster light built by the Schuckert Company, of Nuremberg, Germany, which gives the effect of 816,000,000 candles. An instrument of such power would be useless on board ship, owing to the great amount of current it devours, but in a port, connected with the lighting plant of a large town, it would serve to illumine the country round for many miles.

In addition to its value as an "eye," the search-light can be utilised as an "ear." Ernst Ruhmer, a German scientist, has discovered a method of telephoning along a beam of light from a naval projector. The amount of current passing into the arc is regulated by the pulsations of a telephone battery and transmitter. If the beam be caught by a parabolic reflector, in the focus of which is a selenium cell connected with a battery and a pair of sensitive telephone receivers, the effect of these pulsations of light is *heard*. Selenium being a metal which varies its resistance to an electric circuit in proportion to the intensity of light shining upon it, any fluctuations of the search-light's beams cause electric fluctuations of equal rapidity in the telephone circuit; and since these waves arise from the vibrations of speech, the electric vibrations they cause in the selenium circuit are retransformed at the receiver into the sounds of speech. This German apparatus makes it possible to send messages nine or ten miles over a powerful projector beam.

In the United States Navy, and in other navies as well, night signals are flashed by the electric light. The pattern of lamp used in the United States Navy is divided transversely into two compartments, the upper having a white, the lower a red, lens. Four of these lamps are hung one above the other from a mast. A switch-board connected with the eight incandescent lamps in the series enables the operator to send any required signal, one letter or figure being flashed at a time. During the Spanish-American War the United States fleet made great use of this simple system, which on a clear night is very effective up to distances of four miles.

Large arc-lamps slung on yards over the deck give great help for coaling and unloading vessels at night time. The touch of a switch lights up the deck with the brilliancy of a well-equipped railway station. The day of the "lantern, dimly burning," has long passed away from the big liner, cargo boat, and warship.

WIRELESS TELEGRAPHY INSTRUMENTS

Solitude is being rapidly banished from the earth's surface. By solitude we mean entire separation from news of the world, and the inability to get into touch with people far away. On the remote ranches of the United States, in sequestered Norwegian fiords, in the folds of the eternal hills where the only other living creature is the eagle, man may still be as conversant with what is going on in China or Peru as if he were living in the busy streets of a capital town. The electric wire is the magic news-bringer. Wherever man can go it can go too, and also into many places besides.

We must make one exception—the surface of the sea. Cables rest on ocean's bed, but they would be useless if floated on its surface to act as marine telegraph offices. Winds and waves would soon batter them to pieces, even if they could be moored, which in a thousand fathoms may be considered impracticable.

So until a few years back the occupants of a ship were truly isolated from the time that they left port until they reached land again, except for the rare occasions when a passing vessel might give them a fragment of news.

This has all been changed. Stroll into the saloon of one of our large Atlantic liners and you will see telegram forms lying on the tables. In the 'nineties they would have been about as useful aboard ships as a mackintosh coat in the Sahara. A glance, however, at pamphlets scattered around informs you that the ship carries a Marconi wireless installation, and that a Marconi telegram, handed in at the ship's telegraph office, will be despatched on the wings of ether waves to the land far over the horizon.

Inside the cabin streams of sparks scintillate with a cracking noise, and your message shoots into space from a wire suspended on insulators from one of the mast heads. If circumstances favour, you may receive a reply from the Unseen before the steamer has got out of range of the coast stations. The immense installations at Poldhu, Cornwall, and in Newfoundland, could be used to flash the words to a ship at any point of the transatlantic journey. Owing to lack of space, and consequently power, the steamer's transmitting apparatus has a limited capacity.

The first shipping company to grasp the possibilities of the commercial working of the Marconi system was the Nord-Deutscher-Lloyd, whose mail steamer, *Kaiser Wilhelm der Grosse*, was fitted in March, 1900. At the present time many of the large Atlantic steamship companies carry a wireless installation as a matter of course, ranking it among necessary things. The Cunard, American Atlantic Transport, Allan, Compagnie Transatlantique, Hamburg-American, and Nord-Deutscher-Lloyd lines make full use of the system, as the conveniences it gives far outweigh any expense. A short time since maritime signalling was extremely limited in its range, being effected by flags, semaphores, lights, and sounds, which in stormy weather became uncertain agents, and in foggy, useless. Also the operations of transmitting and receiving were so slow that many a message had to remain uncompleted.

The following paragraph, which appeared in *The Times* of December 11th, 1903, is significant of the very practical value of marine wireless telegraphy. "The American steamer *Kroonland*, from Antwerp for New York, which, as reported yesterday, disabled her steering gear when west of the Fastnet, and had to put back, arrived yesterday morning at Queenstown. The saloon passengers speak in the highest terms of praise of the utility of the Marconi wireless telegraphy with which the liner is fitted, and of the facility with which, when the accident occurred, the passengers were able to communicate with their friends, in England, Scotland, and the Continent, and even America, and get replies before the Irish coast was sighted. The accident occurred on Tuesday about noon, when the liner was 130 miles west of the Fastnet, and communication was at once made with the Marconi station at Crookhaven. Captain Doxrud was enabled accordingly to send messages to the chief agents of the American line, at Antwerp, stating the nature of the damage to the steering gear of the steamer, and that he would have to abandon the idea of prosecuting the western voyage. Within an hour and a half a message was received by the captain from the agents instructing him what to do, and at once the *Kroonland* was headed for Queenstown. Three-fourths of the total number of the saloon passengers and a goodly number of the second cabin sent messages to their friends in various parts of the world, and replies were received even from the Continent before the Fastnet was sighted. Seven or eight passengers telegraphed to relatives for money, and replies were received in four instances, authorising the purser to advance the amounts required, and the money was paid over in each case to the passengers."

The possibility of thus communicating between vessel and land, or vessel and vessel, removes much of the anxiety attending a sea voyage. Business men, for whom even a few days' want of touch with the mercantile markets may be a serious matter, can send long messages in code or otherwise instructing their agents what to do; while they can receive information to shape their actions when they reach land. The "uncommercial traveller" also is pleased and grateful on receiving a message from home. The feeling of loneliness is eliminated. The ocean has lost its right to the term bestowed by Horace—*dissociabilis*, "the separator."

Photo] [Cribb, Southsea.
FIXING A BATTLE-RAM

The ram of a battleship being placed in position with the aid of a huge crane. The size of the ram will be appreciated from the dwarfing effect it has on that of the man perched near the lifting tackle.

Steamship companies vie with one another in their efforts to keep their passengers well posted in the latest news. Bulletins, or small newspapers, are issued daily during the voyage, which give, in very condensed form, accounts of events interesting to those on board. "The amount of fresh news a steamer gathers during a passage is considerable, and is greatly relished by the passengers, who are invariably ravenous for signs of the busy life they left behind, more especially when they have departed on the verge of some important event taking place; and the bulletins are eagerly sought for when it is announced that an inward-bound ship is in communication. The shipowners realise the importance and usefulness of being able to communicate with their commanders before the huge vessels enter narrow waters, and issue instructions concerning their movements.

"The stations, which are placed at carefully-selected points at well-adapted distances around the coast, are connected with either the land telegraph or telephone line, or are close to a telegraph office. They are kept open night and day, as the times of the ships passing are, of course, greatly dependent on the weather encountered during the voyage. For those on shore who are anxious to greet their friends on arrival—with good or bad news, as the case may be—this arrangement enables them to be informed of the exact time of the ship's expected arrival, and they are left free to their own devices, instead of enduring long waits on draughty piers and docks—which, on a wet or windy day, are almost enough to damp the warmest and most enthusiastic welcome.

"Cases have occurred where a telegram, sent from the American side to an outlying English land-station two days after a ship has left, has been transmitted to an outgoing steamer, which in turn has re-transmitted it to the astonished passenger two days prior to his arrival off the English coast; and it has now become quite a common thing for competing teams on vessels many miles apart, and out of sight of each other, to arrange chess matches with each other, some of these interesting events taking two or more days to be played to a finish."[17]

For naval purposes, wireless telegraphy has assumed an importance which can hardly be overestimated, as the whole efficiency of a fine fleet may depend upon a single message flashed through space. All navies are fitting instruments, the British Admiralty being well to the fore. Even in manoeuvres and during the execution of tactical formations the apparatus is constantly at work. The admiral gives the word, and a dozen paper tapes moving jerkily through Morse machines, pass the message round the fleet. The Japanese naval successes have, doubtless, been largely due to their up-to-date employment of this latest development of Western electrical science. No one knows how soon the time may come when the fate of a nation may depend on the proper working of a machine covering a few square feet of a cabin table; for, rapid as has been the growth of wireless telegraphy, it is yet in its infancy.

SAFETY DEVICES

A ship is usually divided into compartments by cross bulkheads of steel. In event of a collision or damage by torpedoes or shell, the water rushing through the break can be prevented from swamping the ship by closing the bulkhead doors.

Messrs. J. Stone and Company, of Deptford, have patented a system of hydraulically operated bulkhead doors, which is finding great favour among shipbuilders on account of its versatility. Each door is closed by an hydraulic cylinder placed above it. The valves of the cylinder are opened automatically by a float when the water rises in the compartment, and every cylinder is also controllable independently from the bridge and other stations in the ship, and by separate hand levers alongside the bulkhead.

The doors can therefore be closed collectively or individually. Should it happen that, when a door has been closed, someone is imprisoned, the prisoner can open the door by depressing a lever inside the compartment, and make his escape. But the door is closed behind him by the action of the float.

THE TRANSMISSION OF POWER ON A SHIP

There are four power agents available on board ship, all derived directly or indirectly from the steam boilers. They are:—

(1) Steam.

(2) High-pressure water.

(3) Compressed air.

(4) Electricity.

On some ships we may find all four working side by side to

drive the multifarious auxiliaries, since each has its peculiar advantages and disadvantages. At the same time, marine engineers prefer to reduce the number as far as possible, since each class of transmission needs specially trained mechanics, and introduces its special complications.

Let us take the four agents in order and briefly consider their value.

Steam is so largely used in all departments of engineering that its working is better understood by the bulk of average mechanics than hydraulic power, compressed air, or electricity. But for marine work it has very serious drawbacks, especially on a war vessel. Imagine a ship which contains a network of steam-pipes running from end to end, and from side to side. The pipes must, on account of the many obstacles they encounter, twist and turn about in a manner which might be avoided on land, where room is more available. Every bend means friction and loss of power. Again, the condensation of steam in long pipes is notorious. Even if they are well jacketed, a great deal of heat will radiate from the ducts into the below-deck atmosphere, which is generally too close and hot to be pleasant without any such further warming. So that, while power is lost, discomfort increases, with a decided lowering of human efficiency. We must not forget, either, the risk attending the presence of a steam-pipe. Were it broken, by accident or in a naval engagement, a great loss of life might result, or, at least, the abandonment of all neighbouring machinery.

For these reasons there is, therefore, a tendency to abolish the direct use of steam in the auxiliary machinery of a modern vessel.

High-pressure water is free from heating and danger troubles, and consequently is used for much heavy work, such as training guns, raising ashes and ammunition, and steering. One of its great advantages is its inelasticity, which prevents the overrunning of gear worked by it. Water, being incompressible, gives a "positive" drive; thus, if the pump delivers a pint at each stroke in the engine-room a pint must pass into the motor, assuming that all joints are tight, and the work due from the passage of one pint is done. Air and steam—and electricity too, if not very delicately controlled—are apt to work in fits and starts when operating against varying resistance, and "run away" from the engineer.

An objection to hydraulic power is, that all leakage from the system must be replaced by fresh water manufactured on board, which, as we have seen, is no easy task.

Compressed air, like steam, may cause explosions; but when it escapes in small quantities only it has a beneficial effect in cooling and freshening the air below decks. The exhaust from an air-driven motor is welcome for the same reason, that it aids ventilation. On a fighting ship it is of the utmost importance that the *personnel* should be in good physical condition; and when the battle-hatches have been battened down for an engagement any supply of fresh oxygen means an increased "staying power" for officers and crew. Poisoned air brings mental slackness, and weakening of resolve; so that if the motive power of heavy machinery can be made to do a second duty, so much the better for all concerned.

Compressed air also proves useful as a water-excluder. If a vessel contain, as it should, a number of water-tight compartments, any water rushing into one of these can be expelled by injecting air until the pressure inside is equal to that of the draught of water of the vessel outside.

On land compressed-air installations include reservoirs of large size in which air can be stored till needed, and which take the place of the accumulator used with hydraulic power. On shipboard want of space reduces such reservoirs to minimum dimensions, so that the compressors must squirt their air almost directly into the cylinders which do the work. When the load, or work, is constantly varying, this direct drive proves somewhat of a nuisance, since the compressors, if worked continuously at their maximum capacity, must waste large quantities of air, while if run spasmodically, as occasion demands, they require much more attention. It is therefore considered advisable by some marine engineers to make compressed air perform as many functions as possible when it is present on a vessel. The United States monitor *Terror* is an instance of a warship which depends on this agency for working her guns and turrets, handling ammunition, and—a somewhat unusual practice—controlling the helm. The last operation is performed by two large cylinders placed face to face athwart the ship. They have a common piston-rod, in the middle of which is a slot for the tiller to pass through. Air is admitted to the cylinders by a valve which is controlled by wires passing over a train of wheels from different stations on the ship. An ingenious device automatically prevents the tiller from moving over too fast, and also helps to lessen the shocks given to the rudder by a heavy sea.

We now come to *electricity*, the fourth and most modern form of transmission. Its chief recommendation is that the wires through which it flows lend themselves readily to a tortuous course without in any way throttling the passage of power. And as every ship must carry a generating plant for lighting purposes, the same staff will serve to tend a second plant for auxiliary machinery. Electric motors work with practically no vibration, are light for their power, and can be very easily controlled from a distance. They therefore enjoy increasing favour; and are found in deck-winches, anchor-capstans, ammunition hoists, ventilation blowers, and cranes. They also control the movements of gun-turrets, having been found most suitable for this work.

If the current were to get loose in a ship it would undoubtedly cause more damage than an escape of compressed air or water. Electricity, even when every known means of keeping it within bounds has been tried, is suspected of causing deterioration to the metalwork of ships. But these disadvantages are not serious enough to hamper the progress of electrical science as applied to marine engineering; and the undoubted economy of the electric motor, its noiselessness, its manageableness, and comparatively small size will, no doubt, in the future lead to its much more extensive use on board our floating palaces and floating forts.

FOOTNOTES:

16. F. M. Bennett, in the Journal of the American Society of Naval Engineers.

17. Charles V. Daly, in *The Magazine of Commerce*.

CHAPTER XIII

"THE NURSE OF THE NAVY"

Just as a navy requires floating distilleries, floating coal stores and floating docks, so does it find very important uses for a floating workshop, which can accompany a fleet to sea and execute such repairs as might otherwise entail the return of a ship to port.

The British Navy has a valuable ally of this kind in the torpedo depôt ship *Vulcan*, which contains so much machinery, in addition to the "auxiliaries" already described, that a short account of this vessel will be interesting.

The *Vulcan*, known as "The Nurse of the Navy," was launched in 1889. She measures 350 feet in length, 58 feet in beam, and has a displacement of 6,830 tons. Her bunkers, of which there are twenty-one, hold 1,000 tons of coal, independently of an extra 300 tons which can be stowed in other neighbouring compartments. When fully coaled she can cruise for 7,000 miles at a speed of 10 knots; or travel at first-class cruiser speed for shorter distances.

The most striking objects on the *Vulcan* are two huge hydraulic cranes, placed almost amidships abreast of one another. They have a total height of 65 feet, and "overhang" 35 feet, so as to be able to lift boats when the torpedo-nets are out and the sides of the vessel cannot be approached. The feet of the cranes sink 30 feet through the ship to secure rigidity, and the upper deck, which bears most of the strain, is strongly reinforced. Inside the pillar of each crane is the lifting machinery, an hydraulic ram $17\frac{1}{2}$ inches in diameter and of 10-foot stroke. By means of fourfold pulleys the lift is increased to 40 feet. When working under the full pressure of 1,000 lbs. to the square inch, the cranes have a hoisting power of twenty tons. In addition to the main ram there is a much smaller one, the function of which is to keep the "slings" (or cables by which the boat is hoisted) taut after a boat has been hooked until the actual moment of lifting comes. But for this arrangement there would be a danger of the slings slackening as the boat rises and falls in a seaway. The small ram controls the larger, and the latter cannot come into action until its auxiliary has tightened up the slings, so that no dangerous jerk can occur when the hoisting begins.

The cranes are revolved by two sets of hydraulic rams, which operate chains passing round drums at the feet of the cranes, and turn them through three-quarters of a circle.

On the *Vulcan's* deck lie six torpedo boats and three despatch boats. The former are 60 feet long, and can attain a speed of 16 knots an hour. When an enemy is sighted these would be sent off to worry the hostile vessels with their deadly torpedoes, and on their return would be quickly picked up and restored to their berths, ready for further use.

The cranes also serve to lift on board heavy pieces of machinery from other vessels for repair.

Photo Cribb.
A 12-inch gun being lowered into its place in the turret of a warship by a gigantic sheer-leg crane, one leg of which is partly visible on the left of the picture.

Down below decks is the workshop, wherein "jobs" are done on the high seas. It has quite a respectable equipment: five lathes, ranging from 15 feet to $3\frac{1}{2}$ feet in length; drilling, planing, slotting, shaping, punching machines; a carpenter's bench; fitters' benches; and a furnace for melting steel. There is also a blacksmith's shop with an hydraulic forging press and a forge blown by machinery; not to mention a large array of tools of all kinds. Special engines are installed to operate the repairs department.

The *Vulcan* also carries search-lights of 25,000 candle-power; bilge pumps which will deliver over 5,000 tons of water per hour; two sets of engines for supplying the hydraulic machinery; air-compressing engines to feed the Whitehead torpedoes; a distilling plant; and last, but by no means least, main engines of 12,000 h.p. drawing steam from four huge cylindrical boilers 17 feet long and 14 feet in diameter.

Altogether, the *Vulcan* is a very complete floating workshop, sufficiently speedy to keep up with a fleet, and even to do scouting work. Her guns and her torpedo craft would render her a very troublesome customer in a fight, though, being practically unarmoured, she would keep as clear of the conflict as possible, acting on the offensive through the proxy of her "hornets." She constitutes the first of a type of vessel which has been suggested by experts, viz. one of high speed and unarmoured, but capable of carrying a swarm of torpedo boats which could be launched in pursuit of the foe. Even if 50 per cent. of the craft were destroyed, the price would be small if a single torpedo were successfully fired at a battleship. The naval motor boat, to which reference has already been made, would just "fill the bill" for such a cruiser; and in the event of a score of them being dropped into the water at a critical moment, they might easily turn the scale in favour of their side.

CHAPTER XIV

THE MECHANISM OF DIVING

DIVING being a profession which can be carried on in its simplest form with the simplest possible apparatus—merely a rope and a stone—its history reaches back into the dim and inexplorable past. We may well believe that the first man who explored the depths of the sea for treasure lived as long ago as the first seeker for minerals in the bosom of the earth. Even when we come to the various appliances which have been gradually developed in the course of centuries, our records are very imperfect. Alexander the Great is said to have descended in a machine which kept him dry, while he sought for fresh worlds to conquer below the waves. Aristotle mentions a device enabling men to remain some time under water. This is all the information, and a very meagre total, too, that we get from classical times.

Stepping across 1,500 years we reach the thirteenth century, about the middle of which Roger Bacon is said to have invented the diving-bell. But like some other discoveries attributed to that Middle-Age physicist, the authenticity of this rests on very slender foundations. In a book published early in the sixteenth century there appears an illustration of a diver wearing a cap or helmet, to which is attached a leather tube floated on the surface of the water by an inflated bag. This is evidently the diving dress in its crudest form; and when we read how, in 1538, two Greeks made a submarine trip under a huge inverted chamber, which kept them dry, in the presence of the great Emperor Charles V. and some 12,000 spectators, we recognise the diving-bell, now so well known.

The latter device did not reach a really practical form till 1717, when Dr. Halley, a member of the Royal Society, built a bell of wood lined with lead. The divers were supplied with air by having casks-full lowered to them as required. To quote his own words: "To supply air to this bell under water, I caused a couple of barrels of about thirty gallons each to be cased with lead, so as to sink empty, each of them having a bunghole in its lowest parts to let in the water, as the air in them condensed on their descent, and to let it out again when they were drawn up full from below. And to a hole in the uppermost parts of these barrels I fixed a leathern hose, long enough to fall below the bunghole, being kept down by a weight appended, so that the air in the upper parts of the barrels could not escape, unless the lower ends of these hose were first lifted up. The air-barrels being thus prepared, I fitted them with tackle proper to make them rise and fall alternately, after the manner of two buckets in a well; and in their descent they were directed by lines fastened to the under edge of the bell, which passed through rings on both sides of the leathern hose in each barrel, so that, sliding down by these lines, they came readily to the hand of a man, who stood on purpose to receive them, and to take up the ends of the hose into the bell. Through these hose, as soon as their ends came above the surface of the water in the barrels, all the air that was included in the upper parts of them was blown with great force into the bell, whilst the water entered at the bungholes below and filled them, and as soon as the air of one barrel had been thus received, upon a signal given that was drawn up, and at the same time the other descended, and by an alternate succession, provided air so quick and in such plenty that I myself have been one of five who have been together at the bottom, in nine to ten fathoms water, for above an hour and a half at a time, without any sort of ill-consequence, and I might have continued there so long as I pleased for anything that appeared to the contrary." After referring to the fact that, when the sea was clear and the sun shining, he could see to read or write in the submerged bell, thanks to a glass window in it, the Doctor goes on to say: "This I take to be an invention applicable to various uses, such as fishing for pearls, diving for coral or sponges and the like, in far greater depths than has hitherto been thought possible; also for the fitting and placing of the foundations of moles, bridges, etc., in rocky bottoms, and for cleaning and scrubbing ships' bottoms when foul, in calm weather at sea. I shall only intimate that, *by an additional contrivance*, I have found it not impracticable for a diver to go out of an engine to a good distance from it, the air being conveyed to him with a continued stream by small flexible pipes, which pipes may serve as a clue to direct him back again when he would return to the bell."

We have italicised certain words to draw attention to the fact that Dr. Halley had invented not only the diving bell, but also the diving dress. Though he foresaw practically all the uses to which diving mechanism could be put, the absence of a means for forcing air *under pressure* into the bell or dress greatly limited the utility of his contrivances, since the deeper they sank below the water the further would the latter rise inside them. It was left for John Smeaton, of Eddystone Lighthouse fame, to introduce the *air-pump* as an auxiliary, which, by making the pressure of the air inside the bell equal to that of the water outside, kept the bell quite free of water. Smeaton replaced Halley's tub by a square, solid cast-iron box, 50 cwt. in weight, large enough to accommodate two men at a time. The modern bell is merely an enlarged edition of this type, furnished with telephones, electric lamps, and, in some cases, with a special air-lock, into which the men may pass when the bell is raised. The pressure in the air-lock is very gradually decreased after the bell has reached the surface, if work has been conducted at great depths, so that the evil effects sometimes attending a sudden change of pressure on the body may be avoided.

Diving bells are very useful for laying submarine masonry, usually consisting of huge stone blocks set in hydraulic cement. Helmet divers explore and prepare the surface on which the blocks are to be placed. Then the bell, slung either from a crane on the masonry already built above water-level, or from a specially fitted barge, comes into action. The block is lowered by its own crane on to the bottom. The bell descends upon it and the crew seize it with tackle suspended inside the bell. Instructions are sent up as to the direction in which the bell should be moved with its burden, and as soon as the exact spot has been reached the signal for lowering is given, and the stone settles on to the cement laid ready for it.

The modern diver is not sent out from a bell, but has his separate and independent apparatus. The first practical diving helmet was that of Kleingert, a German. This enclosed the diver as far as the waist, and constituted a small diving bell, since the bottom was open for the escape of vitiated air. Twenty years later, or just a century after the invention of Halley's bell, Augustus

Siebe, the founder of the present great London firm of Siebe, Gorman, and Company, produced a more convenient "open" dress, consisting of a copper helmet and shoulder-plate in one piece, attached to a waterproof jacket reaching to the hips.

The disadvantage of the open dress was, that the diver had to maintain an almost upright position, or the water would have invaded his helmet. Mr. Siebe therefore added a necessary improvement, and extended the dress to the feet, giving his diver a "close" protection from the water.

We may pass over the gradual development of the "close" dress and glance at the most up-to-date equipment in which the "toilers of the deep" explore the bed of Old Ocean.

The dress—legging, body, and sleeves—is all in one piece, with a large-enough opening at the shoulders for the body to pass through. The helmet, with front and side windows, is attached by a "bayonet joint" to the shoulder-plate, itself made fast to the upper edge of the dress by screws which press a metal ring against the lower edge of the plate so as to pinch the edge of the dress.

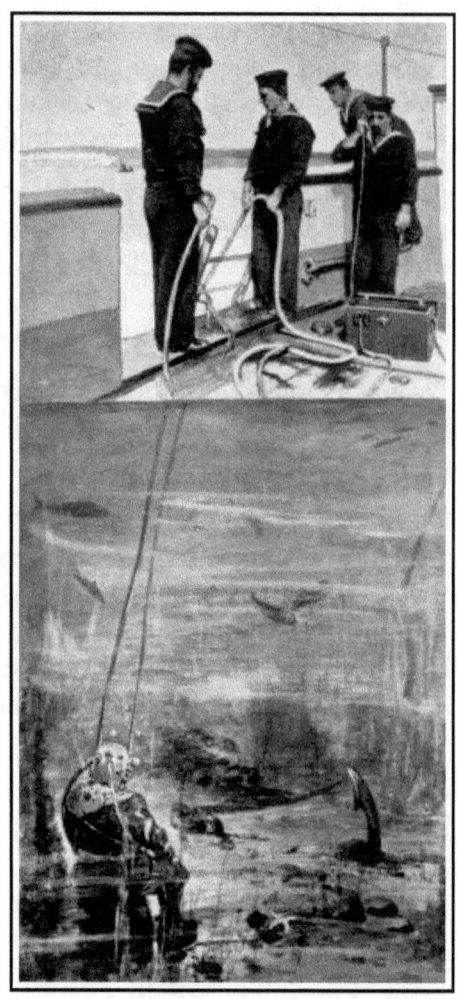

Photo Cribb.
THE DIVER AT WORK

Note the telephone attachment, the wires of which are embedded in the life-line held by the bluejacket on the left. By means of the telephone the diver can give and receive full instructions about his work.

At the back are an inlet and an outlet valve. Between the front and a side window is the transmitter of a loud-sounding telephone, and in the crown the receiver and the button of an electric bell. The telephone wires, and also the wires for a powerful electric light, working on a ball-and-socket joint in front of the dress, are embedded into the life-line. The air-tube, of canvas and rubber, has a stiffening of wire to prevent its being throttled on coming into contact with any object. A pair of weighted boots, each scaling 17 lbs., two 40-lb. lead weights slung over the shoulder, and a knife worn at the waist-belt, complete the outfit of the diver, which, not including the several layers of underclothing necessary to exclude the cold found at great depths, totals nearly 140 lbs. Of this the copper helmet accounts for 36 lbs.

On the surface are the air-pumps, which may be of several types—single-cylinder, double-acting; double-cylinder, double-acting; or three or four cylinder, single-acting—according to the nature of the work. All patterns are so constructed that the valves may be easily removed and examined.

The pressure on a diver increases in the ratio of about $4\frac{1}{4}$ lbs. for every ten feet he descends below the surface. A novice experiences severe pains in the ears and eyes at a few fathoms' depth, which, however, pass off when the pressures both inside and outside of the various organs have become equalised. On rising to the surface again the pains recur, since the external pressure on the body falls more quickly than the internal. The rule for all divers, therefore, is "slow down, slow up." Men of good constitution and resourcefulness are needed for the profession of diving. Only a few can work at extreme depths, though an old hand is able to remain for several hours at a time in sixty feet of water. The record depth reached by a diver is claimed by James Hooper, who, when removing the cargo of the *Cape Horn*, wrecked off the coast of South America, made seven descents to 201 feet, one of which lasted forty-two minutes.

In spite of the dangers and inconveniences attached to his calling, the diver finds in it compensations, and even fascinations, which outweigh its disadvantages. The pay is good—£1 to £2 a day—and in deep-sea salvage he often gets a substantial percentage of all the treasure recovered, the percentage rising as the depth increases. Thus the diver Alexander Lambert, who performed some plucky feats during the driving of the Severn Tunnel,[18] received £4,000 for the recovery of £70,000 worth of gold from the *Alphonso XII.*, sunk off Grand Canary. Divers

Ridyard and Penk recovered £50,000 from the *Hamilla Mitchell*, which lay in 160 feet of water off Shanghai, after nearly being captured by Chinese pirates; and we could add many other instances in which treasure has been rescued from the maw of the sea.

The most useful sphere for a diver is undoubtedly connected with the harbour work and the cleaning of ships' bottoms. For the latter purpose every large warship in the British Navy carries at least one diver. After ships have been long in the water barnacles and marine growths accumulate on the below-water plates in such quantities as to seriously diminish the ship's speed, which means a great waste of fuel, and would entail a loss of efficiency in case of war breaking out. Armed with the proper tools, a gang of divers will soon clean the "foul bottom," at a much smaller cost of time and money than would be incurred by dry-docking the vessel.

The Navy has at Portsmouth, Sheerness, and Devonport schools where diving is taught to picked men, the depth in which they work being gradually increased to 120 feet. Messrs. Siebe and Gorman employ hundreds of divers in all parts of the world, on all kinds of submarine work, and they are able to boast that never has a defect in their apparatus been responsible for a single death. This is due both to the very careful tests to which every article is subjected before it leaves their works, and also to the thorough training given to their employés.

In the sponge and pearl-fishing industries the diving dress is gradually ousting the unaided powers of the naked diver. One man equipped with a standard dress can do the work of twenty natural divers, and do it more efficiently, as he can pick and choose his material.

This chapter may conclude with a reference to the apparatus now used in exploring or rescue work in mines, where deadly fumes have overcome the miners. It consists of an air-tight mask connected by tubes to a chamber full of oxygen and to a bag containing materials which absorb the carbonic acid of exhaled air. The wearer uses the same air over and over again, and is able to remain independent of the outer atmosphere for more than an hour. The apparatus is also useful for firemen when they have to pass through thick smoke.

FOOTNOTE:

18. Vide *The Romance of Modern Engineering*, p. 212.

CHAPTER XV

APPARATUS FOR RAISING SUNKEN SHIPS AND TREASURE

It is somewhat curious that, while the sciences connected with the building of ships have progressed with giant strides, little attention has been paid to the art of raising vessels which have found watery graves in comparatively shallow depths. The total shipping losses of a single year make terrible reading, since they represent the extinction of many brave sailors and the disappearance of huge masses of the world's wealth. A life lost is lost for ever, but cargoes can be recovered if not sunk in water deeper than 180 feet. Yet with all our modern machinery the percentage of vessels raised from even shallow depths is small.

There are practically only two methods of raising a foundered ship: first, to caulk up all leaks and pump her dry; and secondly, to pass cables under her, and lift her bodily by the aid of pontoons, or "camels."

The second method is that more generally used, especially in the estuaries of big rivers where there is a considerable tide. The pontoons, having a united displacement greater than that of the vessel to be raised, are brought over her at low tide. Divers pass under her bottom huge steel cables, which are attached to the "camels." As the tide flows the pontoons sink until they have displaced a weight of water equal to that of the vessel, and then they begin to raise her, and can be towed into shallower water, to repeat the process if necessary next tide. As soon as the deck is above water the vessel may be pumped empty, when all leaks have been stopped.

In water where there is no tide the natural lift must be replaced by artificial power. Under such circumstances the salvage firms use lighters provided with powerful winches, each able to lift up to 800 tons on huge steel cables nearly a foot in diameter. The winches can be moved across a lighter, the cables falling perpendicularly, through transverse wells almost dividing the lighter into separate lengths, so as to get a direct pull. If the wreck has only half the displacement of the lighters, the cables can be passed over rollers on the inner edges of the pontoons, the weight of the raising vessel being counteracted by water let into compartments in the outer side of the pontoons.

There are ten great salvage companies in the British Isles and Europe. The best equipped of these is the Neptune Company, of Stockholm, which has raised 1,500 vessels, worth over £5,000,000 sterling even in their damaged condition, among them the ill-fated submarine "A1." Yet this total represents but a small part of the wealth that has gone to the bottom within a short distance of our coasts.

Turning from the salvage of wrecks to the salvage of precious metal and bulky objects that are known to strew the sea-floor in many places, we must notice the Hydroscope, the invention of Cavaliere Pino, an Italian.

In 1702 there sank in Vigo Bay, on the north-west coast of Spain, twenty-five galleons laden with treasure from America, as the result of an attack by English and Dutch men-of-war. Gold representing £28,000,000 was on those vessels. Down it went to the bottom, and there it is still.

So rich a prize has naturally not failed to attract daring spirits, among whom was Giuseppe Pino. This inventor has produced many devices, the most notable among them the hydroscope, which may best be described as a huge telescope for peering into the depths of the sea. A large circular tank floats on the top of the water. From the centre of its bottom hangs a series of tubes fitting one into the other, so that the whole series can be shortened or lengthened at will. Through the tubes a man can descend to the chamber at their lower extremity, in the sides of which are twelve lenses specially made by Saint Gobain, of Paris, which act as submarine telescopes.

Pino's hydroscope has been at work for some time in Vigo Bay, its operations closely watched by a Spanish war vessel, which will exact 20 per cent. of all treasure recovered. While the hydroscope acts as an eye, the lifting of an object is accomplished by attaching to it large canvas bags furnished with air-tight internal rubber bladders. These have air pumped into them till its pressure overcomes that of the water outside, and the bag then rises like a cork, carrying its load with it. An "elevator"—nine sacks fixed to one frame—will raise twenty-five to thirty tons.

So far Cavaliere Pino has salvaged old Spanish guns, cannon-balls, and pieces of valuable old wood; and presently he may alight on the specie which is the main object of his search.

Another Spanish wreck, the *Florida*, which was a unit of the Spanish Armada, and sank in Tobermory Bay, the Isle of Mull, has many times been attacked by divers. The last attempt made to recover the treasure which that ill-fated vessel was reputed to bear is that of the steam lighter *Sealight*, which employed a very powerful sand pump to suck up any objects which it might encounter on the sea-bottom. Many interesting relics have been raised by the pumps and attendant divers—coins, bones, jewels, timbers, cannon, muskets, pistols, swords, and a compass, which is so constructed that pressure on the top causes the legs to spread. One of the cannon, fifty-four inches long, has a separate powder chamber, the shot and wad still in the gun, and traces of powder in the chamber. It is curious that what we usually consider so modern an invention as the breech-loading cannon should be found side by side with stone balls. The heavier objects were, of course, raised by divers. In this quest also the treasure deposit has not yet been tapped.

CHAPTER XVI

THE HANDLING OF GRAIN

THE ELEVATOR — THE SUCTION PNEUMATIC GRAIN-LIFTER — THE PNEUMATIC BLAST GRAIN-LIFTER — THE COMBINED SYSTEM

THE ELEVATOR

ON or near the quays of our large seaports, London, Liverpool, Manchester, Bristol, Hull, Leith, Dublin, may be seen huge buildings of severe and ugly outline, utterly devoid of any attempt at decoration. Yet we should view them with respect, for they are to the inhabitants of the British Isles what the inland granaries of Egypt were to the dwellers by the Nile in the time of Joseph. Could we strip off the roofs and walls of these structures, we should see vast bins full of wheat, or spacious floors deeply strewn with the material for countless loaves. The grain warehouses of Britain—the Americans would term them "elevators"—have a total capacity of 10,000,000 quarters. Multiply those figures by eight, and you have the number of bushels, each of which will yield the flour for about forty 2-lb. loaves.

In these granaries is stored the grain which comes from abroad. With the opening up of new lands in North and South America, and the exploitation of the great wheat-growing steppes of Russia, English agriculture has declined, and we are content to import five-sixths of our breadstuffs, and an even larger proportion of grain foods for domestic animals. It arrives from the United States, India, Russia, Argentina, Canada, and Australia in vessels often built specially for grain transport; and as it cannot be immediately distributed, must be stored in bulk in properly designed buildings.

These contain either many storeys, over which the grain is spread to get rid of superfluous moisture which might cause dangerous heating; or huge bins, or "silos," in which it can be kept from contact with the air. Experiments have proved that wheat is more successfully preserved if the air is excluded than if left in the open, provided that it is dry. The ancient Egyptians used brick granaries, filled from the top, and tapped at the bottom, in which, to judge by the account of a grievous famine given in the book of Genesis, their wheat was preserved for at least seven years. During last century the silo fell into disrepute; but now we have gone back to the Egyptian plan of closed bins, which are constructed of wood, brick, ferro-concrete, or iron, and are of square, hexagonal, or round section. They are set close together, many under one roof, to economise space; as many as 2,985,000 bushels being provided for in the largest English storehouse.

Such vast quantities of grain require well-devised machinery for their transport from ship to bin or floor, weighing, clearing, and for their transference to barges, coasting vessels, or railway trucks. The Alexander Grain Warehouse of Liverpool may be taken as a typical example of a well-equipped silo granary. It measures 240 by 172 feet, and contains 250 hexagonal bins of brickwork, each 80 feet deep and 12 feet in diameter. The grain is lifted from barges by four elevators placed at intervals along the edge of the quay. The elevator is a wooden case, 40 or 50 feet high, in which an endless band furnished with buckets travels over two rollers placed at the top and bottom. These are let down into the hold and scoop up the grain at the rate of from 75 to 150 tons per hour, according to their size. As soon as a bucket reaches the top roller it empties its charge into a spout, which delivers the grain into a bin, whence it is lifted again 32 feet by a second elevator to a bin from which it flows by gravity to a weighing hopper beneath; and as soon as two tons has collected, the contents are emptied automatically into a distributing hopper. After all this, the grain still has a long journey before it; for it is now shot out on to an endless, flat conveyer belt moving at a rate of 9 to 10 feet per second. It is carried horizontally by this for some distance along the quay, and falls on to a second belt moving at right angles to the first, which whisks it off to the receiving elevators of the storehouse. Once more it is lifted, this time 132 feet, to the top floor of the building, and dropped on to a third belt, which runs over a movable throwing-off carriage. This can be placed at any point of the belt's travel, to transfer the grain to any of the spouts leading to the 250 bins.

Here it rests for a time. When needed for the market it flows out at the bottom of a bin on to belts leading to delivery elevators, from which it may be either passed back to a storage bin after being well aired, or shot into wagons or vessels. From first to last a single grain may have to travel three miles between the ship and the truck without being touched once by a human hand.

The vertical transport of grain is generally effected by an endless belt, to which buckets are attached at short intervals. The grain, fed to the buckets either by hand or by mechanical means, is scooped up, whirled aloft, and when it has passed the topmost point of its travel, and just as the bucket is commencing the descent, it flies by centrifugal force into a hopper which guides it to the travelling belt, as already described.

Of late years, however, much attention has been paid to pneumatic methods of elevating, by which a cargo is transferred from ship to storehouse, or from ship to ship, through flexible tubes, the motive power being either the pressure of atmospheric air rushing in to fill a vacuum, or high-pressure air which blows the grain through the tube in much the same way as a steam injector forces water into a boiler. Sometimes both systems are used in combination. We will first consider these methods separately.

THE SUCTION PNEUMATIC GRAIN-LIFTER

is the invention of Mr. Fred E. Duckham, engineer of the Millwall Docks, London. The ships in which grain is brought to England often contain a "mixed" cargo as well; and that the unloading of this may proceed simultaneously with the moving of the wheat it is necessary to keep the hatches clear. As long as the grain is directly under a hatchway, a bucket elevator can reach it; but all that is not so conveniently situated must be brought within range of the buckets. This means a large bill for labour, even if machinery is employed to help the "trimming." Mr. Duckham therefore designed an elevator which could easily reach any corner of a ship's interior. The principal parts are a large cylindrical air-tight tank, an engine to exhaust air from the same, and long hoses, armoured inside with a steel lining, connected at one end to the tank, and furnished at the other with a nozzle. These hoses extend from the receiving tank to the grain, which, when the air has been exhausted to five or six pounds to the square inch, flies up the tubes into the tank. At the bottom of the tank are ingenious air-locks, to allow the grain to pass into a bin below without admitting air to spoil the vacuum. The locks are automatic, and as soon as a certain quantity of grain has collected, tip sideways, closing the port through which it flowed, and allowing it to drop through a hinged door. Two locks are attached together, the one discharging while the other is filling. An elevator of this kind will shift 150 tons or more an hour. Mr. Duckham claims for his invention that it has no limit in capacity. It is practically independent of everything but its own steam power; and the labour of one man suffices to keep its flexible suckers buried in grain. No corner is inaccessible to the nozzle. The pipes occupy only a very small part of the hatchway. They can be set to work immediately a vessel comes alongside. As many as a quarter of a million bushels are handled daily by one of these machines.

The pneumatic elevator is often installed on a floating base, so that it may be moved about in a dock.

THE PNEUMATIC BLAST GRAIN-LIFTER

differs from the system just described in that the grain is *driven* through the pipes or hoses by air compressed to several pounds above atmospheric pressure. A small tube attached to the main hose conveys compressed air to the nozzle through which grain enters the tube. The nozzle consists of a short length of metal piping which is buried in the grain. One half of it is encased by a jacket into which the compressed air rushes. As the air escapes at high speed past the inner end of the piping into the main hose, it causes a vacuum in the piping and draws in grain, which is shot up the hose by the pressure behind it. As already remarked, the action of this pneumatic elevator is similar to that of a steam injector.

THE COMBINED SYSTEM

Under some conditions it is found convenient to employ both

suction and blast in combination: suction to draw the grain from a vessel's hold into elevators, from which it is transferred to the warehouse by blast. Special boats are built for this work, *e.g.* the *Garryowen*, which has on board suction plant for transferring grain from a ship to barges, and also blowing apparatus for elevating it into storehouses or into another ship. The *Garryowen* has the hull and engines of an ordinary screw steamer, so that it can ply up and down the Shannon and partly unload a vessel to reduce its draught sufficiently to allow it to reach Limerick Docks. Floating elevators of this kind are able to handle upwards of 150 tons of grain per hour.

CHAPTER XVII

MECHANICAL TRANSPORTERS AND CONVEYERS

MECHANICAL CONVEYERS — ROPEWAYS — CABLEWAYS —
TELPHERAGE — COALING WARSHIPS AT SEA

A MAN carrying a sack of coal over a plank laid from the wharf to the ship's side, a bricklayer's labourer moving slowly up a ladder with his hod of mortar—these illustrate the most primitive methods of shifting material from one spot to another. When the wheelbarrow is used in the one case, and a rope and pulley in the other, an advance has been made, but the effort is still great in proportion to the work accomplished; and were such processes universal in the great industries connected with mining and manufacture, the labour bill would be ruinous.

The development of methods of transportation has gone on simultaneously with the improvement of machinery of all kinds. To be successful, an industry must be conducted economically throughout. Thus, to follow the history of wheat from the time that it is selected for sowing till it forms a loaf, we see it mechanically placed in the ground, mechanically reaped, threshed, and dressed, mechanically hauled to the elevator, mechanically transferred to the bins of the same, mechanically shot into trucks or a ship, mechanically raised into a flour-mill, where it is cleaned, ground, weighed, packed, and trucked by machinery, mechanically mixed with yeast and baked, and possibly distributed by mechanically operated vehicles. As a result we get a 2-lb. loaf for less than three-pence. Anyone who thinks that the price is regulated merely by the *amount* of wheat grown is greatly mistaken, for the cheapness of handling and transportation conduces at least equally to the cheapness of the finished article.

The same may be said of the metal articles with which every house is furnished. A fender would be dearer than it is were not the iron ore cheaply transported from mine to rail, from rail to the smelting furnace, from the ground to the top of the furnace. In short, to whatever industry we look, in which large quantities of raw or finished material have to be moved, stored, and distributed, the mechanical conveyer has supplanted human labour to such an extent that in lack of such devices we can scarcely conceive how the industry could be conducted without either proving ruinous to the people who control it or enhancing prices enormously.

The types of elevators and conveyers now commonly used in all parts of the world are so numerous that in the following pages only some selected examples can be treated.

Speaking broadly, the mechanical transporter can be classified under two main heads—(1) those which handle materials *continuously*, as in the case of belt conveyers, pneumatic grain dischargers, etc.; and (2) those which work *intermittently*, such as the telpher, which carries skips on an aerial ropeway. The first class are most useful for short distances; the latter for longer distances, or where the conditions are such that the material must be transported in large masses at a time by powerful grabs.

Some transporters work only in a vertical direction; others only horizontally; while a third large section combine the two movements. Again, while some are mere conveyers of material shot into or attached to them, others scoop up their loads as they move. The distinctions in detail are numerous, and will be brought out in the chapters devoted to the various types.

MECHANICAL CONVEYERS

We have already noticed band conveyers in connection with the transportation of grain. They are also used for handling coal, coke, diamond "dirt," gold ore, and other minerals, and for moving filled sacks. The belts are sometimes made of rubber or of balata faced with rubber on the upper surface, which has to stand most of the wear and tear—sometimes of metal plates joined together by hinges at the ends.

A modification of the belt is the continuous trough, with sloping or vertical sides. This is built of open-ended sections jointed so that they may pass round the terminal rollers. While travelling in a straight line the sides of the sections touch, preventing any escape of the material carried, but at the rollers the ends open in a

V-shape.

Another form of conveyer has a stationary trough through which the substance to be handled is pulled along by plates attached to cables or endless chains running on rollers. Or the moving agency may be plates dragged backwards and forwards periodically, the plates hanging in one direction only, like flap valves, so as to pass over the material during the backward stroke, and bite it during the forward stroke. The vibrating conveyer is a trough which moves bodily backwards and forwards on hinged supports, the oscillation gradually shaking its contents along. As no dragging or pushing plates are here needed, this form of conveyer is very suitable for materials which are liable to be injured by rough treatment.

ROPEWAYS

A certain person on asking what was the distance from X to Y, received the reply, "It is ten miles as the crow flies." The country being mountainous, the answer did not satisfy him, and he said, "Oh! but you see, I am *not* a crow." Engineers laying out a railway can sympathise with this gentleman, for they know from sad experience that places only a few miles apart in a straight line often require a track many miles long to connect them if gradients are to be kept moderate.

Now a locomotive, a railway carriage, or a goods truck is very heavy, and must run on the firm bosom of Mother Earth. But for comparatively light bodies a path may be made which much more nearly resembles the proverbial flight of the crow, or, as our American cousins would say, a bee-line. If you have travelled in Norway and Switzerland you probably have noticed here and there steel wire ropes spanning a torrent or hanging across a narrow valley. Over these ropes the peasants shoot their hay crops or wood faggots from the mountain-side to their homes, or to a point near a road where the material can be transferred to carts. Adventurous folk even dare to entrust their own bodies to the seemingly frail steel thread, using a brake to control the velocity of the descent.

The history of the modern ropeway and cableway dates from

the 'thirties, when the invention of wire rope supplied a flexible carrying agent of great strength in proportion to its weight, and of sufficient hardness to resist much wear and tear, and too inelastic to stretch under repeated stresses. To prevent confusion, we may at once state that a ropeway is an aerial track used only for the *conveyance* of material; whereas a cableway hoists as well as conveys. A further distinction—though it does not hold good in all cases—may be seen in the fact that, while cableways are of a single span, ropeways are carried for distances ranging up to twenty miles over towers or poles placed at convenient intervals.

Ropeways fall into two main classes: first, those in which the rope supporting the weight of the thing carried moves; secondly, those in which the carrier rope is stationary, and the skips, or tubs, etc., are dragged along it by a second rope. The moving rope system is best adapted for light loads, not exceeding six hundredweight or so; but over the second class bodies scaling five or six tons have often been moved. In both systems the line may be single or double, according to the amount of traffic which it has to accommodate. The chief advantage of the double ropeway is that it permits a continuous service and an economy of power, since in cases where material has to be delivered at a lower level than the point at which it is shipped, the weight of the descending full trucks can be utilised to haul up ascending empty trucks. Spans of 2,000 feet or two-fifths of a mile are not at all unusual in very rough country where the spots on which supports can be erected are few and far between; but engineers naturally endeavour to make the span as short as possible, in order to be able to use a small size of rope.

Glancing at some interesting ropeways, we may first notice that used in the construction of the new Beachy Head Lighthouse, recently erected on the foreshore below the head on which the original structure stands. For the sake of convenience, the workshops, storage yards, etc., were placed on the cliffs, 400 feet above the sea and some 800 feet in a direct line from the site of the new lighthouse. Between the cliff summit and a staging in the sea were stretched two huge steel ropes, the one, six inches in circumference, for the track over which the four-ton blocks of granite used in the building, machinery, tools, etc., should be lowered; the other, $5\frac{1}{2}$ inches in circumference, for the return of the carriers and trucks containing workmen. The ropes had a breaking strain of 120 and 100 tons respectively; that is to say, if put in an hydraulic testing machine they would have withstood pulls equal to those exerted by masses of these weights hung on them. Their top ends were anchored in solid rock; their lower ends to a mass of concrete built up in the chalk forming the sea-bottom. When a granite block was attached to the carrier travelling on the rope, its weight was gradually transferred to the rope by lowering the truck on which it had arrived until the latter was clear of the block. As soon as the stone started on its journey the truck was lifted again to the level of the rails and trundled away. A brakesman, stationed at a point whence he could command the whole ropeway, had under his hand the brake wheels regulating the movements of the trailing ropes for lowering and hauling on the two tracks.

Another interesting ropeway is that at Hong-Kong, which transports the workmen in a sugar factory on the low, fever-breeding levels to their homes in the hills where they may sleep secure from noxious microbes. The carriers accommodate six men at a time, and move at the rate of eight miles an hour. The sensation of being hauled through mid-air must be an exhilarating one, and some of us would not mind changing places with the workmen for a trip or two, reassured by the fact that this ropeway has been in operation for several years without any accident.

In Southern India, in the Anamalai Hills, a ropeway is used for delivering sawn timber from the forests to a point $1\frac{1}{4}$ miles below. Prior to the establishment of this ropeway the logs were sent down a circuitous mountain track on bullock carts. Its erection was a matter of great difficulty, on account of the steep gradients and the dense and unhealthy forest through which a path had to be cut; not to mention the dragging uphill of a cable which, with the reel on which it was wound, weighed four tons. For this last operation the combined strength of nine elephants and a number of coolies had to be requisitioned, since the friction of the rope dragging on the ground was enormous. However, the engineers soon had the cable stretched over its supports, and the winding machinery in place at the top of the grade. The single rope serves for both up and down traffic; a central crossing station being provided at which the descending can pass the ascending carrier. Seven sleepers at a time are sent flying down the track at a rate of twenty miles an hour: a load departing every half-hour. The saving of labour, time, and expense is said to be very great, and when the saw mills have a larger output the economy of working will be still more remarkable.

The longest passenger ropeway ever built is probably that over the Chilkoot Pass in Alaska, which was constructed in 1897 and 1898 to transport miners from Dyea to Crater Lake on their way to the Yukon goldfields. From Crater Lake to the Klondike the Yukon River serves as a natural road, but the climb to its head waters was a matter of great difficulty, especially during the winter months, and accompanied by much suffering. But when the trestles had been erected for the fixed ropes, two in number, miners and their kits were hauled over the seven miles at little physical cost, though naturally the charges for transportation ruled higher than in less rugged regions. The opening of the White Pass Railway from Skagway has largely abolished the need for this cable track, which has nevertheless done very useful work. The Chilkoot ropeway has at least two spans of over 1,500 feet. As an engineering enterprise it claims our consideration, since the conveyance of ropes, timber, engines, etc., into so inhospitable a region, and the piecing of them together, demanded great persistence on the part of the engineers and their employés.

CABLEWAYS

For removing the "over-burden" of surface mines and dumping it in suitable places, for excavating canals, for dredging, and for many other operations in which matter has to be moved comparatively short distances, the cableway is largely employed. We have already noticed that it differs from the ropeway in that it has to hoist and discharge its burdens as well as convey them.

The cableway generally consists of a single span between two towers, which are either fixed or movable on rails according to the requirements of the work to be done. In addition to the main cable which bears the weight, and the rope which moves the skips along it, the cableway has the "fall" rope, which lowers the skip to the ground and raises it; the dumping rope, which discharges it; and the "button" rope, which pulls blocks off the horn of the skip truck at intervals as the latter moves, to support the "fall" rope from the main cable. If the fall rope sagged its weight would, after a certain amount had been paid out, overcome the weight of the skip, and render it impossible to lower the skip to the filling point. So a series of fall-rope carriers are, at the commencement of a journey from one end of the cableway, riding on an arm in front of the skip carriage. The button-rope, passing under a pulley on the top of the skip carriage, is furnished at intervals with buttons of a size increasing towards the point at which the skip must be lowered. The holes in the carriers are similarly graduated so as to pass over any button but the one intended to arrest them. If we watched a skip travelling to the lowering point, we should notice that the carriers were successively pulled off the skip carriage by the buttons, and strung along over the main cable and under the fall rope.

When the skip has been lowered and filled the fall and hauling ropes are wound in; the skip rises to the main cable, and begins to travel towards the dumping point. As long as the dumping rope is also hauled in at the same rate as the hauling rope it has no effect on the skip, but when its rate of travel is increased by moving it on to a larger winding drum, the skip is tipped or opened, as the case may be, without being arrested.

The skip may be filled by hand or made self-filling where circumstances permit.

The cableway is so economical in its working that it has greatly advanced the process of "open-pit" mining. Where ore lies near the surface it is desirable to remove the useless overlying matter (called "over-burden") bodily, and to convey it right away, in preference to sinking shallow shafts with their attendant drawbacks of timbering and pumping. An inclined railway is handicapped by the fact that it must occupy some of the surface to be uncovered, while liable to blockage by the débris of blasting operations. The suspended cableway neither obstructs anything nor can be obstructed, and is profitably employed when a ton of ore is laid bare for every four tons of over-burden removed. In the case of the Tilly Foster Mine, New York, where the removal of 300,000 tons of rock exposed 600,000 tons of ore from an excavation 450 ft. long by 300 ft. wide, the saving effected by the cableway was enormous. Again, referring to the Chicago Drainage Canal, "the records show that while labourers, sledging and filling into cars, averaged only 7 to $8\frac{1}{2}$ cubic yards per man per day, in filling into skips for the cable ways the labourers averaged from 12 to 17 cubic yards per day."[19] The first cableway erected by the Lidgerwood Manufacturing Company for the prosecution of this engineering work handled 10,821 cubic yards a month, and proved so successful that nineteen similar plants were added. The cableways are suspended in this instance from two towers moving on parallel tracks on each bank of the canal, the towers being heavily ballasted on the outer sides of their bases to counteract the pull of the cable. From time to time, when a length had been cleared, the towers were moved forward by engines hauling on fixed anchors.

The cableway is much used in the erection of masonry piers for bridges across rivers or valleys. Materials are conveyed by it rapidly and easily to points over the piers and lowered into position. Spans of over 1,500 feet have been exceeded for such purposes; and if need be, spans of 2,000 feet could be made to carry loads of twenty-five tons at a rate of twenty miles an hour.

TELPHERAGE

On most ropeways the skips or other conveyances are moved along the fixed ropes by trailing ropes working round drums driven by steam and controlled by brakes. But the employment of electricity has provided a system called *telpherage*, in which the vehicle carries its own motor, fed by current from the rope on which it runs and from auxiliary cables suspended a short distance above the main rope. "Telpher" is a term derived from two Greek words signifying "a far carrier," since the motor so named will move any distance so long as a track and current is supplied to it. The carrier—for ore, coal, earth, barrels, sacks, timber, etc.—is suspended from the telpher by the usual hook-shaped support common to ropeways, to enable the load to pass the arms of the posts or trestles bearing the rope. The telpher usually has two motors, one placed on each side of a two-wheeled carriage so as to balance; but sometimes only a single motor is employed. Just above the running cable is the "trolley" cable, from which the telpher picks up current through a hinged arm, after the manner of an electric tram. The carriers are controlled on steep grades by an electric braking device, which acts automatically, its effect varying with the speed at which the telpher runs. The carrier wheels, driven by the motors, adhere to the cable without slipping on grades as severe as three in ten, even when the surface has been moistened by rain. "In order to stop the telpher at any desired point, the trolley wire is divided into a number of sections, each controlled by a switch conveniently located. By opening a switch the current is cut off from the corresponding section, and the telpher will stop when it reaches this point. It is again started by closing the switch. At curves a section of the trolley wire (*i.e.* overhead cable for current) is connected to the source of current through a 'resistance' which lowers the voltage (pressure of the current) across the motors at this point. Thus, upon approaching a curve, the telpher automatically slows down, runs slowly around the curve until it passes the resistance section, and is then automatically accelerated."[20]

The telpher line is very useful (for transporting material considerable distances) in districts where it would not pay to construct a surface railway. On plantations it serves admirably to shift grain, fruits, tobacco, and other agricultural products. Then, again, a wide field is open to it for transmitting light articles, such as castings and parts of machinery, from one part of a foundry or manufactory to another, or from factory to vessel or truck for shipment. When coal has to be handled, the buckets are dumped automatically into bins.

The telpher has much the same advantages over the steam-worked ropeway that an electric tram has over one moved by an endless cable. Its control is easier; there is less friction; and the speed is higher. And in common with ropeways it can claim independence of obstructions on the ground, and the ability to cross ravines with ease, which in the case of a railway would have to be bridged at great expense.

COALING WARSHIPS AT SEA

The war between Russia and Japan has brought prominently before the public the necessity of being able to keep a war vessel well supplied with coal: a task by no means easy when coaling stations are few and far between. The voyage of Admiral Rojdestvensky from Russia to Eastern waters was marked by occasions on which he entered neutral ports to draw supplies for his furnaces, though we know that colliers sailed with the warships to replenish their exhausted bunkers. In the old days of sailing vessels, their motive power, even if fitful, was inexhaustible. But now that steam reigns supreme as the mover of the world's floating forts, the problem of "keeping the sea" has become in one way very much more complicated. The radius of a vessel's action is limited by the capacity of her coal bunkers. Her captain in war time would be perpetually perplexed by the question of fuel, since movement is essential to naval success, while any misjudged fast steaming in pursuit of the enemy might render his ship an inert mass, incapable of motion, because the coal supplies had given out; or at least might compel him to return for supplies to the nearest port at a slow speed, losing valuable time.

A TEMPERLEY-MILLER MARINE CABLEWAY COALING H.M.S. "TRAFALGAR" AT SEA

A carrier, from which are slung the sacks of coal, is hauled backwards and forwards by steel ropes stretching between the foremast of the transport and a mast rigged on the warship.

Just as a competitor in a long-distance race takes his nourishment without halting, so should a battleship be able to coal "on the wing." The task of transferring so many tons of the mineral from one ship's hold to that of another may seem easy enough to the inexperienced critic, and under favourable conditions it might not be attended by great difficulty. "Why," someone may say, "you have only to bring the collier alongside the warship, make her fast, and heave out the coals." In a perfect calm this might be feasible; but let the slightest swell arise, and then how the sides of the two craft would bump together, with dire results to the weaker party! Actual tests have shown this.

At present "broadside" coaling is considered impracticable, but the "from bow to stern" method has passed through its initial stages, and after many failures has reached a point of considerable efficiency. The difficulties in transferring coal from a collier to a warship by which she is being towed will be apparent after very little reflection. In the first place, there is the danger of the cableway and its load dipping into the water, should the distance between the two vessels be suddenly diminished, and the corresponding danger of the cable snapping should the pitching of the vessels increase the distance between the terminals of the cableway. These difficulties have made it impossible to merely shoot coals down a rope attached high up a mast of the collier and to the deck of the warship. What is evidently needed is some system which shall pay the cableway out or take it in automatically, so as to counterbalance any lengthening or shortening movement of the vessels.

The Lidgerwood Manufacturing Company of New York, under the direction of Mr. Spencer Miller, have brought out a cableway specially adapted for marine work. The two vessels concerned are attached by a stout tow-line, the collier, of course, being in the rear. To carry the load, a single endless wire rope, $\frac{3}{8}$ inch in diameter and 2,000 feet long, is employed. It spans the distance between collier and ship twice, giving an inward track for full sacks, and an outward track for their return to the collier. On one vessel are two winches, the drums of which both turn in the same direction; but while one drum is rigidly attached to its axle, the other slips under a stress greater than that needed to keep the rope sufficiently taut. Since the rope passes round a pulley at the other terminal, pressure placed at any point on the rope will tend to tighten both tracks, while a slackening at any point would similarly ease them. Supposing, then, that the ships suddenly approach, there will be a certain amount of slack at once wound in; if, on the other hand, the ships draw apart, the slipping drum will pay out rope sufficient to supply the need. The constant slipping of this drum sets up great heat, which is dissipated by currents of air. As the sacks of coal arrive on the man-of-war they are automatically detached from the cable, and fall down a chute into the hold.

In the Temperley Miller Marine Cableway the load is carried on a main cable kept taut by a friction drum, and the hauling is done by an endless rope which has its own separate winches. In actual tests made at sea in rough weather sixty tons per hour have been transferred, the vessels moving at from four to eight miles an hour.

FOOTNOTES:

19. *Cassier's Magazine.*

20. *Cassier's Magazine.*

CHAPTER XVIII

AUTOMATIC WEIGHERS

SCARCELY less important than the rapid transference of materials from one place to another is the quick and accurate weighing of the same. If a pneumatic grain elevator were used in conjunction with an ordinary set of scales such as are to be found at a corn dealer's there would be great delay, and the advantage of the elevator would largely be lost. Similarly a mechanical transporter of coal or ore should automatically register the tonnage of the mineral handled, to prevent undue waste of time.

There are in existence many types of automatic weighing machines, the general principles of which vary with the nature of the commodity to be weighed. Finely divided substances, such as grain, seeds, and sugar, are usually handled by *hopper* weighers. The grain, etc., is passed into a bin, from the bottom of which it flows into a large pan. When the proper unit of weight—a hundredweight or a ton—has nearly been attained, the flow is automatically throttled, so that it may be more exactly controlled, and as soon as the full amount has passed, the machine closes the hopper door and tips the pan over. The latter delivers its contents and returns to its original position, while the door above is simultaneously opened for the operation to be repeated. A counting apparatus records the number of tips, so that a glance suffices to learn how much material has passed through the weigher, which may be locked up and allowed to look after itself for hours together. The "Chronos" automatic grain scale is built in many sizes for charges of from 12 to 3,300 lbs. of grain, and tips five times a minute. Avery's grain weigher takes up to $5\frac{1}{2}$ tons at a time.

For materials of a lumpy nature, such as coal and ore, a different method is generally used. The hopper process would not be absolutely accurate, since the rate of feed cannot be exactly controlled when dust and large lumps weighing half a hundredweight or more are all jumbled together. Therefore instead of a pan which tips automatically as soon as it has received a fixed weight, we find a bin which, when a quantity roughly equal to the correct amount has been let in, sinks on to a weigher and has its contents registered by an automatic counter, which continuously adds up the total of a number of weighings and displays it on a dial. So that if there be 10 lbs. in excess of a ton at the first charge, the dial records "one ton," and keeps the 10 lbs. "up its sleeve" against the next weighing, to which the excess is added. Avery's mineral scale works, however, on much the same principle as that for grain already noticed, a special device being fitted to render the feed to the weighing pan as regular as possible. His weigher is used to feed mechanical furnace stokers. The quantity of coal used can thus be checked, while an automatic apparatus prevents the stoker bunkers from being overfilled.

Continuous weighers register the amount carried by a conveyer while in motion. The recording apparatus comes into action at fixed intervals, *e.g.* as soon as the conveyer has moved ten feet. The weighing mechanism is practically part of the conveyer, and takes the weight of ten feet. The steelyard is adjusted to exactly counterbalance the unloaded belt or skips of its length, but rises in proportion to the load. As soon as the conveyer has travelled ten feet the weight on the machine is immediately recorded, and the steelyard returns to zero.

Intermittent weighers record the weight of trucks or tubs passing over a railway or the cables of aerial track, the weigher forming part of the track and coming into play as soon as a load is fully on it.

Some machines not only weigh material, but also stow and pack it. We find a good instance in Timewell's sacking apparatus, which weighs corn, chaff, flour, oatmeal, rice, coffee, etc., transfers it to sacks, and *sews the sack up* automatically. The amount of time saved by such a machine must be very great.

NOTE.—The author desires to express his indebtedness to Mr. George F. Zimmer's *The Mechanical Handling of Material* for some of the information contained in the above chapter; and to the publishers, Messrs. A. Crosby Lockwood and Son, for permission to make use of the same.

CHAPTER XIX

TRANSPORTER BRIDGES

When the writer was in Rouen, in 1898, two lofty iron towers were being constructed by the Seine: the one on the Quai du Havre, the other on the Quai Capelier, which borders the river on the side of the suburb St. Sever.

The towers rose so far towards the sky that one had to throw one's head very far back to watch the workmen perched on the summit of the framework. What were the towers for? They seemed much too slender for the piers of an ordinary suspension bridge fit to carry heavy traffic. An inquiry produced the information that they were the first instalment of a "transbordeur," or transporter bridge. What is a bridge of this kind?

Well, it may best be described as a very lofty suspension bridge, the girder of which is far above the water to allow the passage of masted ships. The suspended girder serves only as the run-way for a truck from which a travelling car hangs by stout steel ropes, the bottom of the car being but a few feet above the water. The truck is carried across from tower to tower, either by electric motors or by cables operated by steam-power.

The transporter bridge in a primitive form has existed for some centuries, but its present design is of very modern growth. With the increase of population has come an increased need for uninterrupted communication. Where rivers intervene they must be bridged, and we see a steady growth in the number of bridges in London, Paris, New York, and other large towns.

Unfortunately a bridge, while joining land to land, separates water from water, and the dislocation of river traffic might not be compensated by the conveniences given to land traffic. The Forth, Brooklyn, Saltash, and other bridges have, therefore, been built of such a height as to leave sufficient head-room under the girders for the masts of the tallest ships.

But what money they have cost! And even the Tower Bridge, with its hinged bascules, or leaves, and bridges with centres revolving horizontally, devour large sums.

Wanted, therefore, an efficient means of transport across a river which, though not costly to install, shall offer a good service and not impede river traffic.

Thirty years ago Mr. Charles Smith, a Hartlepool engineer, designed a bridge of the transporter type for crossing the Tees at Middlesbrough. The bridge was not built, because people feared that the towers would not stand the buffets of the north-easterly gales.

The idea promulgated by an Englishman was taken up by foreign engineers, who have erected bridges in Spain, Tunis, and France. So successful has this type of ferry-bridge proved, that it is now receiving recognition in the land of its birth, and at the present time transporter bridges are nearing completion in Wales and on the Mersey.

THE LATEST TYPE OF BRIDGE

The Transporter Bridge at Bizerta, Tunis. It has a span of 500 feet, and the suspension girder is 120 feet above high water, so that the largest vessels may pass under it from the Mediterranean to the inland lakes. The car is seen near the bottom of the right-hand tower.

The first "transbordeur" built was that spanning the Nervion, a river flowing into the Bay of Biscay near Bilbao, a Spanish town famous for the great deposits of iron ore close by. A pair of towers rises on each bank to a height of 240 feet, and carry a suspended trussed girder 530 feet long at a level of 150 feet above high-water mark. The car, giving accommodation for 200 passengers (it does not handle vehicles), hangs on the end of cables 130 feet long, and is propelled by a steam-engine situated in one of the towers. Motion is controlled by the car-conductor, who is connected electrically with the engine-room. The lofty towers are supported on the landward side by stout steel ropes firmly anchored in the ground. These ropes are carried over the girder in the familiar curve of the suspension bridge, and attached to it at regular intervals by vertical steel braces. The cost of the bridge—£32,000—compares favourably with that of any alternative non-traffic-blocking scheme, and the graceful, airy lines of the erection are by no means a blot on the landscape.

The second "transbordeur" is that of Rouen, already referred to. Its span is rather less—467 feet—but the suspension girder lies higher by 14 feet. The car is 42 feet long by 36 broad, and weighs, with a full load, 60 tons. A passage, which occupies 55 seconds, costs one penny first class, one halfpenny second class; while a vehicle and horses pay $2\frac{1}{2}$ d. to 4d., according to weight. The car is propelled by electricity, under the control of a man in the conning-tower perched on the roof.

At Bizerta we find the third flying-ferry, which connects that town with Tunis, over a narrow channel between the Mediterranean Sea and two inland lakes. It replaced a steam-ferry which had done duty for about ten years.

The lakes being an anchorage for war vessels, it was imperative that any bridge over the straits should not interrupt free ingress and egress. This bridge has a span of 500 feet, and like that at Bilbao is worked by steam. Light as the structure appears, it has withstood a cyclone which did great damage in the neighbourhood. It is reported that the French Government has decided to remove the bridge to some other port, because its prominence would make it serve as a range-finder for an enemy's cannon in time of war. Its place would be taken either by a floating-bridge or by a submarine tunnel.

The Nantes "transporter" over the Loire differs from its fellows in one respect, viz. that it is built on the cantilever or balance principle. Instead of a single girder spanning the space between the towers, it has three girders, the two end ones being balanced on the towers and anchored at their landward extremities by vertical cables. The gap between them is bridged by a third girder of bow shape, which is stiff enough in itself to need no central support. The motive power is electricity.

All these structures will soon be eclipsed by two English bridges: the one over the Usk at Newport, Monmouthshire; the other over the Mersey and Manchester Ship Canal at Runcorn "Gap," where the river narrows to 1,200 feet.

The first of these has towers 250 feet high and 685 feet apart. The girders will give 170 feet head-room above high-water mark. Five hundred passengers will be able to travel at one time on the car, besides a number of road vehicles, and as the passage is calculated to take only one minute, the average velocity will exceed eight miles an hour. The cost has been set down at £65,000, or about one-thirtieth that of a suspension bridge, and one-third that of a bascule bridge. The bridge is being built by the French engineers responsible for the Rouen *transbordeur*.

Coming to the much more imposing Runcorn bridge we find even these figures exceeded. This span is 1,000 feet in length. The designer, Mr. John J. Webster, has already made a name with the Great Wheel which, at Earl's Court, London, has given many thousands of pleasure-seekers an aerial trip above the roofs of the metropolis. The following account by Mr. W. G. Archer in the *Magazine of Commerce* describes this mammoth of its kind in some detail:—

"The two main towers carrying the cables and the stiffening girders are built, one on the south side of the Ship Canal, and the other on the foreshore on the north bank of the river; and the approaches consist of new roadways, nearly flat, built between stone and concrete retaining walls as far as the water's edge, and a corrugated steel flooring, upon which are laid the timber blocks on concrete, resting on steel elliptical girders and cast-iron columns. The roadway in front of the towers is widened out to 70 feet, for marshalling the traffic, and for providing space for waiting-rooms, etc. The towers are constructed wholly of steel, rise 190 feet above high-water level, and are bolted firmly to the cast-iron cylinders below. Each tower consists of four legs, spaced 30 feet apart at the base, and each pair of towers are 70 feet apart, and are braced together with strong horizontal and diagonal frames. Each of the two main cables consists of 19 steel ropes bound together, each rope being built up of 127 wires 0·16 inches in diameter. The ends of the cable backstays are anchored into the solid rock on each side of the river, about 30 feet from the rock surface. The weight of the main cables is about 243 tons, and from them are suspended two longitudinal stiffening girders, 18 feet deep, and placed 35 feet apart horizontally, the underside of the girders being 82 feet above the level of high water.... Upon the lower flange of the stiffening girders are fixed the rails upon which runs the traveller, from which is suspended the car. The traveller is 77 feet long, and is carried by sixteen wheels on each rail. It is propelled by two electric motors of about 35 horse-power each.... The car will be capable of holding at one time four large wagons and 300 passengers, the latter being protected from the weather by a glazed shelter.... The time occupied by the car in crossing will be $2\frac{1}{4}$ minutes, so, allowing for the time spent in loading and unloading, it will be capable of making nine or ten trips per hour. This bridge, when completed, will have the largest span of any bridge in the United Kingdom designed for carrying road traffic, the clear space over the Mersey and Ship Canal being 1,000 feet.... The total cost of the structure, including Parliamentary expenses, will be about £150,000."

Mr. Archer adds that, in spite of prophecies of disastrous collisions between transporter cars and passing ships, there has up to date been no accident of any kind. To those in search of a new sensation the experience of skimming swiftly a few feet above the water may be recommended.

CHAPTER XX

BOAT AND SHIP RAISING LIFTS

In modern locomotion, whether by land or water, it becomes increasingly necessary to keep the way unobstructed where traffic is confined to the narrow limits of a pair of rails, a road, or a canal channel. We widen our roads; we double and quadruple our rails. Canals are, as a rule, not alterable except at immense cost; and if, in the first instance, they were not built broad enough for the work that they are afterwards called upon to do, much of their business must pass to rival methods of transportation. Modern canals, such as the Manchester and Kiel canals, were given generous proportions to start with, as their purpose was to pass ocean-going ships, and for many years it will not be necessary to enlarge them. The Suez Canal has been widened in recent years, by means of dredgers, which easily scoop out the sandy soil through which it runs and deposit it on the banks. But the Corinth Canal, cut through solid rock, cannot be thus economically expanded, and as a result it has proved a commercial failure.

Even if a canal be of full capacity in its channel-way there are points at which its traffic is throttled. However gently the country it traverses may slope, there must occur at intervals the necessity of making a lock for transferring vessels from one level to the other. Sometimes the ascent or descent is effected by a series of steps, or flight of locks, on account of the magnitude of the fall; and in such cases the loss of time becomes a serious addition to the cost of transport.

In several instances engineers have got over the difficulty by ingenious hydraulic lifts, which in a few minutes pass a boat through a perpendicular distance of many feet. At Anderton, where the Trent and Mersey Canal meets the Weaver Navigation, barges up to 100 tons displacement are raised fifty feet. Two troughs, each weighing with their contents 240 tons, are carried by two cast-iron rams placed under their centres, the cylinders of which are connected by piping. When both troughs are full the pressure on the rams is equal, and no movement results; but if six inches of water be transferred from the one to the other, the heavier at once forces up the lighter. At Fontinettes, on the Neufosse Canal, in France, at La Louvière, in Belgium, and at Peterborough, in Canada, similar installations are found; the last handling vessels of 400 tons through a rise of 65 feet.

Fine engineering feats as these are, they do not equal the canal-lift on the Dortmund-Ems Canal, which puts Dortmund in direct water communication with the Elbe, and opens the coal and iron deposits of the Rhine and Upper Silesia to the busy manufacturing district lying between these two localities. About ten miles from its eastern extremity the main reach of the canal forks off at Heinrichenburg, from the northward branch running to Dortmund, its level being on the average some 49 feet lower than the branch. For the transference of boats an "up" and "down" line of four locks each would have been needed; and apart from the inevitable two hours' delay for locking, this method would have entailed the loss of a great quantity of precious water.

Mr. R. Gerdau, a prominent engineer of Düsseldorf-Grafenburg, therefore suggested an hydraulic lift, which should accommodate boats of 700 tons, and pass them from the one level to the other in five minutes.

This scheme was approved, and has recently been completed. The principle of the lift is as follows:—A trough, 233 feet long, rests on five vertical supports, themselves carried by as many hollow cylindrical floats moving up and down in deep wells full of water. The buoyancy of the five floats is just equal to the combined weight of the trough and its load, so that a comparatively small force causes the latter to rise or fall, as required. By letting off water from the trough—which is, of course, furnished with doors to seal its ends—it would be made to ascend; while the addition of a few tons would cause a descent. But this would mean waste of water; and, were the trough not otherwise governed, a serious accident might happen if a float sprang a leak. Motion is therefore imparted to the trough by four huge vertical screws, resting on solid masonry piers, and turning in large collars attached to the trough near its corners. All the screws work in unison through gearing, as they are sufficiently stout to bear the whole load; even were the floats removed, no tilting or sudden fall is possible. The screws are driven by an electric motor of 150 horse-power, perched on the girders joining the tops of four steel towers which act as guides for the trough to move in, while they absorb all wind-pressure. Under normal circumstances the trough rises or sinks at a speed of four inches per second. The total mass in motion—trough, water, boat, and floats—is 3,100 tons. Our ideas of a float do not ordinarily rise above the small cork which we take with us when we go a-fishing, or at the most above the buoy which bobs up and down to mark a fair-way. These five "floats"—so called—belong to a very much larger class of creations. Each is 30 feet across inside and $46\frac{1}{2}$ feet high. Their wells, 138 feet deep, are lined with concrete nearly a yard thick, to ensure absolute water-tightness, inside the stout iron casings, which rise 82 feet above the bottom.

In view of the immense weight which they have to carry, the piers under the screw-spindles are extremely solid. At its base each measures 14 feet by 12 feet 4 inches, and tapers upwards for 36 feet till these dimensions have contracted to 8 feet 10 inches by 6 feet 6 inches. The spindles, 80 feet long and 11 inches in diameter, must be four of the largest screws in existence. To make it absolutely certain that they contained no flaws, a 4-inch central hole was drilled through them longitudinally—another considerable workshop feat. If shafts of such length were left unsupported when the trough was at its highest point, there would be danger of their bending and breaking; and they are, therefore, provided with four sliding collars each, connected each to its fellow by a rod. When the trough has risen a fifth of its travel the first rod lifts the first collar, which moves in the guide-pillars. This in turn raises the second; the second the third; and so on. So that by the time the trough is fully raised each spindle is kept in line by four intermediate supports.

The trough, 233 feet long by $34\frac{1}{2}$ feet wide, will receive a vessel 223 feet long between perpendiculars. It has a rectangular section, and is built up of stout plates laid on strong cross-girders, all carried by a single huge longitudinal girder resting on the float columns.

One of the most difficult problems inseparable from a structure of this kind is the provision of a water-tight joint between the trough and the upper and lower reaches of the canal. At each end of the trough is a sliding door faced on its outer edges with indiarubber, which the pressure of the water inside holds tightly against flanges when pressure on the outside is removed. The termination of the canal reaches have similar doors; but as it would be impossible to arrange things so accurately that the two sets of flanges should be water-tight, a wedge, shaped like a big **U**, and faced on both sides with rubber, is interposed. The wedge at the lower reach gate is thickest at the bottom; the upper wedge the reverse; so that the trough in both cases jams it tight as it comes to rest. The wedges can be raised or lowered in accordance with the fluctuations of the canals.

After thus briefly outlining the main constructional features of the lift, let us watch a boat pass through from the lower to the upper level. It is a steamer of 600 tons burden, quite a formidable craft to meet so far inland; while some distance away it blows a warning whistle, and the motor-man at his post moves a lever which sets the screw in motion. The trough sinks until it has reached the proper level, when the current is automatically broken, and it sinks no further. Its travel is thus controllable to within $\frac{3}{16}$ of an inch.

An interlocking arrangement makes it impossible to open the trough or reach gates until the trough has settled or risen to the level of the water outside. On the other hand, the motor driving the lifting screws cannot be started until the gates have been closed, so that an accidental flooding of the countryside is amply provided against.

A man now turns the crank of a winch on the canal bank and unlocks the canal gate. A second twist couples the gates between the canal and the trough together and starts the lifting-motors overhead, which raise the twenty-eight ton mass twenty-three feet clear of the water-level. The boat enters; the doors are lowered and uncoupled; the reach gate is locked. The spindle-motor now starts; up "she" goes, and the process of coupling and raising gates is repeated before she is released into the upper reach. From start to finish the transfer occupies about five minutes.

If a boat is not self-propelled, electric capstans help it to enter and leave the trough. Such a vessel could not be passed through in less than twenty minutes.

Putting on one side the ship dry docks, which can raise a 15,000 ton vessel clear of the sea, the Dortmund hydraulic lift is the largest lift in the world, and the novelty of its design will, it is hoped, render the above account acceptable to the reader. Before leaving the subject another canal lift may be noticed—that on the Grand Junction Canal at Foxton, Leicestershire—which has replaced a system of ten locks, to raise barges through a height of 75 feet.

The new method is the invention of Messrs. G. and C. B. J. Thomas. In principle it consists of an inclined railway, having eight rails, four for the "up" and as many for the "down" traffic. On each set of four rails runs a tank mounted on eight wheels, which is connected with a similar tank on the other set by 7-inch steel-wire ropes passing round winding drums at the top of the incline. The tanks are thus balanced. At the foot of the incline a barge which has to ascend is floated into whichever tank may be ready to receive it, and the end gate is closed. An engine is then started, and the laden tank slides "broadside on" up the 300-foot slope. The summit being reached, the tank gates are brought into register with those of the upper reach, and as soon as they have been opened the boat floats out into the upper canal. Boats of 70 tons can be thus transferred in about twelve minutes, at a cost of but a few pence each. On a busy day 6,000 tons are handled.

By permission of] [Mr. Gordon Thomas.
A BOAT LIFT

A canal barge lift which has superseded ten locks at Foxton, Leicestershire. Two tanks, balancing one another, run on separate tracks up and down an incline. At the bottom and top of the incline the tank is submerged so that a barge may float in or out.

A SHIP-RAISING LIFT

The writer has treated one form of lift for raising ships out of the water—the floating dry dock—elsewhere,[21] so his remarks in this place will be confined to mechanism which, having its foundations on Mother Earth, heaves mighty vessels out of their proper element by the force of hydraulic pressure. Looking round for a good example of an hydraulic ship-lift, we select that of the Union Ironworks, San Francisco.

Some years ago the works were moved from the heart of the city to the edge of Mission Bay, with the object of carrying on a large business in marine engineering and shipbuilding. For such a purpose a dry dock, which in a short time will lift a vessel clear of the water for cleaning or repairs, is of great importance to both owners and workmen. By the courtesy of the proprietors of *Cassier's Magazine* we are allowed to append the following account of this interesting lift.

The site available for a dock at the Union Ironworks was a mud-flat. The depth of soft mud being from 80 to 90 feet, would render the working of a graving dock (*i.e.* one dug out of the ground and pumped dry when the entrance doors have been closed) very disagreeable; as such docks, where much mud is carried in with the water, require a long time to be cleaned and to dry out. Plans were therefore prepared by Mr. George W. Dickie for an hydraulic dock, including an automatic control, which the designer felt confident would meet all the requirements of the situation, and which, after careful consideration, the Union Ironworks decided to build. Work was begun in January, 1886, and the dock was opened for business on June 15th, 1887—a very fine record.

This dock consists of a platform built of cross and longitudinal steel girders, 62 feet wide and 440 feet long, having keel blocks and sliding bilge blocks upon which the ship to be lifted rests. The lifting power is generated by a set of four steam-driven, single-acting horizontal plunger pumps, the diameter of the plungers being $3\frac{1}{2}$ inches and the stroke 36 inches. Forty strokes per minute is the regular speed.

There is a weighted accumulator, or regulator, connected with the pumps, the throttle valve of the engines being controlled by the accumulator.[22] The load on the accumulator consists of a number of flat discs of metal, the first one about 14 inches thick and the others about 4 inches thick, the diameter being about 4 feet. The first disc gives a pressure of 300 lbs. per square inch. This is sufficient to lift the dock platform without a ship, and is always kept on.

In lifting a ship, as she comes out of the water and gets heavier on the platform, additional discs are taken on by the accumulator ram as required. The discs are suspended by pins on the side catching into links of a chain. The engineer, to take on another disc, unhooks the throttle from the accumulator rod, runs the engine a little above the normal speed, the accumulator rises and takes the weight of the disc to be added; the link carrying that disc is thus relieved and is withdrawn. The engineer again hooks the accumulator rod to the engine throttle, and the whole is self-acting again until another weight is required. When all the discs are on the ram the full pressure of 1,100 lbs. per square inch is reached, which enables a ship of 4,000 tons weight to be raised.

There are eighteen hydraulic rams on each side of the dock. These rams are each 30 inches in diameter and have a stroke of 16 feet; and as the platform rises 2 feet for 1 foot movement of the rams, the total vertical movement of the platform is 32 feet. When lowered to the lowest limit there are 22 feet of water over the keel blocks at high tide.

The foundations consist of seventy-two cylinders of iron, which extend from the top girders to several feet below the mud line. These cylinders are driven full of piles, no pile being shorter than 90 feet. The cylinders are to protect the piles from the *teredo* (the timber-boring worm), which is very destructive in San Francisco Harbour. A heavy cast-iron cap completes each of the foundation piers, and two heavy steel girders extend the full length of the dock on each side, resting on the foundation piers and uniting them all longitudinally. The hydraulic cylinders are carried by large castings resting on the girders, each having a central opening to receive a cylinder, which passes down between the piers. There are thirty-six foundation piers, and eighteen hydraulic cylinders on each side of the dock.

On the top of each hydraulic ram is a heavy sheave or pulley, 6 feet in diameter, over which pass eight steel cables, 2 inches in diameter, making in all 288 cables. One end of each cable is anchored in the bed-plates supporting the hydraulic cylinders, while the other end is secured to the side girders of the platform. Each of the cables has been tested with a load of 80 tons, so that the total test load for the ropes has been 21,000 tons.

In lifting a ship the load is never evenly distributed on the platform. There is, in fact, often more than one ship on the platform at once. Some rams, therefore, may have a full load and others much less. Under these conditions, to keep the platform a true plane, irrespective of the irregular distribution of the load, Mr. Dickie designed a special valve gear to make the action of the dock perfectly automatic. Down each side of the dock a shaft is carried, operated by a special engine in the power house. At each hydraulic ram this shaft carries a worm, gearing with a worm-wheel on a vertical screw extending the full height reached by the stroke of the ram. This screw works in a nut on the end of a lever, the other end of which is attached to the ram. Between the two points of support a rod, working the valves—also carried by the ram—engages with the lever. If at a given moment the screw-end is raised, say, six inches, the lever opens the valve. As the ram rises, the lever, having its other end similarly lifted by the rise, gradually assumes a horizontal position, and the valve closes.

To lift the dock the engine working the valve shaft is started, and with it the operating screws. These, through the levers, open the inlet valves. The rams now begin to move up: if any one has a light load it will move up ahead of the other, but in doing so it lifts the other end of the lever and closes the valve. In fact, the screws are continually opening the valves, while the motion of the rams is continually closing them, so that no ram can move ahead of its screw, and the speed of the screw determines the rate of movement of the lifting platform.

To lower the dock, the engine operating the valve shaft is reversed, and the screws and levers then control the outlet valves as they controlled the inlet valves in raising. When the platform has reached the limit of its movement, a line of locks on top of the foundation girders, thirty-six on each side, are pushed under the platform by an hydraulic cylinder, and the platform is lowered on to them, where it rests until the work is done on the ship; then the platform is again lifted, the locks are drawn back, and the platform with its load is lowered until the ship floats out. All the operations are automatic.

Since the dock was opened well over a thousand ships have been lifted in it without any accident whatever; the total register tonnage approaching 2,000,000. The great favour in which the dock is held by shipowners and captains is partly due to the fact already mentioned, that the ship is lifted above the level of tide water, where the air can circulate freely under the bottom, thus quickly taking up all the moisture, and where the workmen can carry on operations with greater comfort.

When extensive repairs have to be undertaken on iron or steel vessels, the fact that this dock forms part of an extensive shipbuilding plant, and is located right in the yard, enables such repairs to be executed with despatch and economy. Several large steamships have had the under-water portions of their hulls practically rebuilt in this dock. The steamship *Columbia*, of the Oregon Line, had practically a new bottom, including the whole of the keel, completed in twenty-six days. This is possible, because every facility is alongside the dock and the bottom of the vessel is on a level with the yard.

This being the only hydraulic dock controlled automatically (in 1897), it has attracted a large amount of attention from engineering experts in this class of work. English, French, German, and Russian engineers have visited the Union Iron Works to study its working, and their reports have done much to bring the facilities offered to shipping for repairs by the Union Iron Works to the notice of shipowners all the world over.

FOOTNOTES:

21. *The Romance of Modern Engineering*, pp. 383 foll.

22. For explanation of the "accumulator," see the chapter on Hydraulic Tools (p. 81).

CHAPTER XXI

A SELF-MOVING STAIRCASE

At the American Exhibition, held in the Crystal Palace in 1902, there was shown a staircase which, on payment of a penny, transported any sufficiently daring person from the ground-floor to the gallery above. All that the experimenters had to do was to step boldly on, take hold of the balustrade, which moved at an equal pace with the stairs, and step off when the upper level was reached.

The "escalator" (Latin *scalae* = flight of stairs) hails from the United States, where it is proving a serious rival to the elevator. In principle, it is a continuously working lift, the slow travel of which is more than compensated by the fact that it is always available. The ordinary elevator is very useful in a large business or commercial house, where it saves the legs of people who, if they had to tramp up flight after flight of stairs, would probably not spend so much money as they would be ready to part with if their vertical travel from one floor to another was entirely free of effort. But the ordinary lift is, like a railway, intermittent. We all know what it means to stand at the grille and watch the cage slide downwards on its journey of, perhaps, four floors, when we want to go to a floor higher up. Rather than face the delay we use our legs.

Theoretically, therefore, a large emporium should contain at least two lifts. If the number be further increased, the would-be passenger will have a still better chance of getting off at once. Thus at the station of the Central London Railway we have to wait but a very few seconds before a grille is thrown back and an attendant invites us to "Hurry up there, please!"

Yet there is delay while the cage is being filled. The actual journey occupies but a small fraction of the time which elapses between the moment when the first passenger enters the lift at the one end of the trip and the moment when the last person leaves it at the other end. In a building where the lift stops every fifteen feet or so to take people on or put them off, the waste of time is still more accentuated.

The escalator is always ready. You step on and are transported one stage. A second staircase takes you on at once if you desire it. There is no delay. Furthermore, the room occupied by a single escalator is much less than that occupied by the number of lifts required to give anything like an equally efficient service.

In large American stores, then, it is coming into favour, and also on the Manhattan Elevated Railway of New York. When once the little nervousness accompanying the first use has worn off, it eclipses the lift. A writer in *Cassier's Magazine* says: "In one large retail store during the holiday season more than 6,000 persons per hour have been carried upon the escalator for five hours of the day, and the aggregate for an entire day is believed to be 50,000. In the same store on an ordinary day the passengers alighting at the second floor from the eight large lifts, which run from the basement to the fifth floor, were counted, likewise the number at the escalator. This latter was found to be 859 per cent. of the number delivered by the eight lifts. In another establishment, in a very busy hour, the number taken from the first floor by the escalator was four times the number taken from the first floor by the fourteen lifts, which were running at their maximum capacity. To the merchant this spells opportunity for business.

"The experience at the Twenty-third Street and Sixth Avenue station of the Manhattan Elevated Railway in New York, during a recent shut-down of the escalator, which has been in service for some time, is interesting as showing the attitude of the public, of which many millions have been carried by the installation during the several years of its operation. The daily traffic receipts of this station for a period beginning several weeks before the shut-down and extending as many after, for the years 1903 and 1902, and receipts of the adjacent stations for the same period were carefully plotted ... and the loss area during the period of shut-down was determined. The loss area was found to embrace 64,645 fares. It was, furthermore, daily a matter of observation that numbers of people, finding that the escalator was not running, refused to climb the stairs, and turned away from the station.

"In the case of a great store, the escalator may be constructed as one continuous machine, with landings at each floor, and so arranged that steps which carry passengers up may perform a like service in carrying others down; or separate machines may be installed in various locations affording the best opportunity for displaying merchandise to the customer who may be proceeding from the lower to the upper floor. In the case of a six-storey building so equipped with escalator service in both directions, or in all ten escalator flights, it is obvious that the facilities are equal to an impossible number of elevators; and as facility of access has a direct bearing upon opportunities for business, it may well be argued that the relative value, measured by rent, of the main and upper floors is greatly changed."

Each step in a staircase has two parts—the "tread" or horizontal board on which the foot is placed, and the vertical "riser" which acts both as a support to the tread above and also prevents the foot from slipping under the tread. In the escalator each tread is attached rigidly to its riser, and the two together form an independent unit.

For the convenience of passengers in stepping on or off at the upper and lower landings, the treads in these places are all in the same horizontal plane. As they approach the incline the risers gradually appear, and the treads separate vertically. At the top of the incline the process is gradually reversed, the risers disappearing until the treads once more form a horizontal belt.

The means of effecting this change is most ingenious. Each tread and its riser is carried on a couple of vertical triangular brackets, one at each side of the staircase. The base of the bracket is uppermost, to engage with the tread, and its apex has a hole through which passes a transverse bar, which in its central part forms a pin in the link-chain by which power is transmitted to the escalator. Naturally, the step would tip over. This is prevented by a yoke attached to each end of the bar, at right angles to it and parallel to the tread. The yoke has at each extremity a small wheel running on its own rail—there being two rails for each side of the staircase.

Since step, brackets, bar, and yoke are all rigidly joined

together, the step is unable to leave the horizontal, but its relation to the steps above and below is determined by the arrangement of the rails on which the yoke wheels run. When these are in the same plane, all the yokes, and consequently the treads, will also be in the same plane. But at the incline, where the inner rail gradually sinks lower than its fellow, the front wheel of one tread is lower than the front wheel of the next, and the risers appear. It may be added that, owing to the double track at each side of the staircase, the back wheel of one tread does not interfere with the front wheel of that below; and that on the level they come abreast without jostling, as the yoke is bent.

The chain, of which the step-bars form pins, travels under the centre of the staircase. It is made up of links eighteen inches long, having, in addition to the bars, a number of steel cross-pins $1\frac{1}{2}$ inches in diameter, their axes three inches apart, so that the chain as a whole has a three-inch "pitch." The hubs of the links are bushed with bronze, and have a graphite "inlay," which makes them self-lubricating. Every joint is turned to within $\frac{1}{1,000}$ inch of absolute accuracy.

The tracks are of steel and hardwood, insulated from the ironwork which supports them by sheets of rubber. The wheels are so constructed as to be practically noiseless, so that as a whole the escalator works very quietly.

"It has been observed," says the authority already quoted, "that beginners take pains to step upon a single tread, and that after a little experience no attention whatever is given to the footing, owing to the facility of adapting oneself to the situation. The upper landing is somewhat longer, thereby affording an interval for stepping off at either side of sufficient duration to meet the requirements of the aged and infirm. The sole function of the travelling landing is to provide a time interval to meet the requirements of the slowest-acting passenger, and not of the alert. The terminal of the exit landing, be it top or bottom (for the escalator operates equally well for either ascent or descent), is a barrier, called the shunt, of which the lower member travels horizontally in a plane oblique to the direction of movement of the steps, and at a speed proportionately greater, thereby imparting a right-angle resultant to the person or obstacle on the step which may come in contact with the shunt. By reason of this resultant motion, the person or obstacle is gently pushed off the end of the step upon the floor, without shock or injury in the slightest degree. The motion of the escalator is so smooth and constant that it does not interpose the least obstacle to the free movement of the passenger, who may walk in either direction or assume any attitude to the same degree as upon a stationary staircase."

At Cleveland, U.S.A., there has been erected a rolling roadway, consisting of an inclined endless belt and platform made of planks eight feet long, placed transversely across the roadway. The timbers are fastened together in trucks of two planks each, adjoining trucks being joined by heavy links to form a moving roadway, which runs on 4,000 small wheels. At each end the roadway, which is continuous, passes round enormous rollers. Its total length is 420 feet, and the rise 65 feet. Four electric motors placed at regular intervals along its length, and all controlled by one man at the head of the incline, drive it at three miles an hour. It can accommodate six wagons at a time.

CHAPTER XXII

PNEUMATIC MAIL TUBES

You put your money on the counter. The shop assistant makes out a bill; and you wonder what he will do with it next. These large stores know nothing of an open till. Yet there are no cashiers' desks visible; nor any overhead wires to whisk a carrier off to some corner where a young lady, enthroned in a box, controls all the pecuniary affairs of that department.

While you are wondering the assistant has wrapped the coin in the bill and put the two into a dumb-bell-shaped carrier, which he drops into a hole. A few seconds later, flop! and the carrier has returned into a basket under another opening. There is something so mysterious about the operation that you ask questions, and it is explained to you that there are pneumatic tubes running from every counter in the building to a central pay-desk on the first or second floor; and that an engine somewhere in the basement is hard at work all day compressing air to shoot the carriers through their tubes.

Certainly a great improvement on those croquet-ball receptacles which progressed with a deliberation maddening to anyone in a hurry along a wooden suspended railway! Now, imagine tubes of this sort, only of much larger diameter, in some cases, passing for miles under the streets and houses, and you will have an idea of what the Pneumatic Mail Despatch means: the cash and bill being replaced by letters, telegrams, and possibly small parcels.

"Swift as the wind" is a phrase often in our mouths, when we wish to emphasise the celerity of an individual, an animal, or a machine in getting from one spot of the earth's surface to another. Mercury, the messenger of uncertain-tempered Jove, was pictured with wings on his feet to convey, symbolically, the same notion of speed. The modern human messenger is so poor a counterpart of the god, and his feet are so far from being winged, that for certain purposes we have fallen back on elemental air-currents, not unrestrained like the breezes, but confined to the narrow and certain paths of the metal tube.

The pneumatic despatch, which at the present day is by no means universal, has been tried in various forms for several decades. Its first public installation dates from 1853, when a tube three inches in diameter and 220 yards long was laid in London to connect the International Telegraph Company with the Stock Exchange. A vacuum was created artificially in front of the carrier, which the ordinary pressure of the atmosphere forced through the tube. Soon after this the post-office authorities took the matter up, as the pneumatic system promised to be useful for the transmission of letters; but refused to face the initial expense of laying the tube lines.

When, in 1858, Mr. C. F. Varley introduced the high pressure method, pneumatic despatch received an impetus comparable to that given to the steam-engine by the employment of high-pressure steam. It was now possible to use a double line of tubes economically, the air compressed for sending the carriers through the one line being pumped out of a chamber which sucked them back through the other. Tubes for postal work were soon installed in many large towns in Great Britain, Europe, and the United States; including the thirty-inch pneumatic railway between the North-Western District post office in Eversholt Street and Euston Station, which for some months of 1863 transported the mails between these two points. The air was exhausted in front of the carriage by a large fan. Encouraged by its success, the company built a much larger tube, nearly $4\frac{1}{2}$ feet in diameter, to connect Euston Station with the General Post Office. This carried fourteen tons of post-office matter from one end to the other in a quarter of an hour. There was an intermediate station in Holborn, where the engines for exhausting had been installed. But owing to the difficulty of preventing air leakage round the carriages the undertaking proved a commercial failure, and for years the very route of this pneumatic railway could not be found; so quickly are "failures" forgotten!

The more useful small tube grew most vigorously in America and France. In, or about, the year 1875 the Western Union Telegraph Company laid tubes in New York to despatch telegrams from one part of the city to the other, because they found it quicker to send them this way than over the wires. Eighteen years later fifteen miles of tubes were installed in Chicago to connect the main offices of the same company with the newspaper offices in the town, and with various important public buildings. Messages which formerly took an hour or more in delivery are now flipped from end to end in a few seconds.

The Philadelphia people meanwhile had been busy with a double line of six-inch tubes, 3,000 feet long, laid by Mr. B. C. Batcheller between the Bourse and the General Post Office, for the carriage of mails. The first thing to pass through was a Bible wrapped in the "Stars and Stripes." A 30 horse-power engine is kept busy exhausting and compressing the air needed for the service, which amounts to about 800 cubic feet per minute. Philadelphia can also boast an eight-inch service, connecting the General Post Office with the Union Railway Station, a mile away. One and a half minutes suffice for the transit of the large carriers packed tightly with letters and circulars, nearly half a million of which are handled by these tubes daily.

New York is equally well served. Tubes run from the General Post Office to the Produce Exchange, to Brooklyn, and to the Grand Central Station. The last is $3\frac{1}{2}$ miles distant; but seven minutes only are needed for a tube journey which formerly occupied the mail vans for nearly three-quarters of an hour.

Paris is the city of the *petit bleu*, so important an institution in the gay capital. Here a network of tubes connects every post office in the urban area with a central bureau, acting the part of a telephone exchange. If you want to send an express message to a friend anywhere in Paris, you buy a *petit bleu*, *i.e.* a very thin letter-card not exceeding $\frac{1}{4}$ oz. in weight, at the nearest post office, and post it in a special box. It whirls away to the exchange, and is delivered from there if its destination be close at hand; otherwise it makes a second journey to the office most conveniently situated for delivery. Everybody uses the *voie pneumatique* of Paris, so much cheaper than, and quite as expeditious as, the telegraph; with the additional advantage that all messages are transmitted in the sender's own handwriting. The system has been instituted for a quarter of a century, and the Parisians would feel lost without it.

London is by no means tubeless, for it has over forty miles of $1\frac{1}{2}$, $2\frac{1}{4}$, and 3-inch lines radiating from the postal nerve-centre of the metropolis, of lengths ranging from 100 to 2,000 yards. The tubes are in all cases composed of lead, enclosed in a protecting iron piping. To make a joint great care must be exercised, so as to avoid any irregularity of bore. When a length of piping is added to the line, a chain is first passed through it, which has at the end a bright steel mandrel just a shade larger than the pipe's internal diameter. This is heated and pushed half-way into the pipe already laid; and the new length is forced on to the other half till the ends touch. A plumber's joint having been made, the mandrel is drawn by the chain through the new length, obliterating any dents or malformations in the interior.

The main lines are doubled—an "up" and a "down" track; short branches have one tube only to work the inward and the outward despatches.

The carriers are made of gutta-percha covered with felt. One end is closed by felt discs fitting the tube accurately to prevent the passage of air, the other is open for the introduction of messages. As they fly through the tube, the carriers work an automatic signalling apparatus, which tells how far they have progressed and when it will be safe to despatch the next carrier.

The London post-office system is worked by six large engines situated in the basement of the General Post Office.

So useful has the pneumatic tube proved that a Bill has been before Parliament for supplying London with a 12-inch network of tubes, totalling 100 miles of double line. In a letter published in *The Times*, April 19, 1905, the promoters of the scheme give a succinct account of their intentions, and of the benefits which they expect to accrue from the scheme if brought to completion. The Batcheller system, they write, with which it is proposed to equip London, is not a development of the miniature systems used for telegrams or single letters here or in Paris, Berlin, and other cities. Such systems deal with a felt carrier weighing a few ounces, which is stopped by being blown into a box. The

Batcheller system deals with a loaded steel carrier weighing seventy pounds travelling with a very high momentum. The difference is fundamental. In this sense pneumatic tubes are a recent invention, and absolutely new to Europe.

The Batcheller system is the response to a pressing need. Careful observations show that more than 30 per cent. of the street traffic is occupied with parcels and mails. These form a distinct class, differentiated from passengers on the one hand and from heavy goods on the other. The Batcheller system will do for parcels and mails what the underground electric railways do for passengers. It has been in use for twelve years in America for mail purposes, and where used has come to be regarded as indispensable.

The plan for London provides for nearly one hundred miles of double tubes with about twice that number of stations for receiving and delivery. The system will cover practically the County of London, and no point within that area can be more than one-quarter of a mile from a tube station. Beyond the County of London deliveries will be made by a carefully organised suburban motor-cart service. Thirty of the receiving stations are to be established in the large stores. The diameter of the tube is to be of a size that will accommodate 80 per cent. of the parcels, as now wrapped, and 90 per cent. with slight adaptation. The remaining 10 per cent.—furniture, pianos, and other heavy goods —are to be dealt with by a supplementary motor service. If the tubes were enlarged their object would be partially defeated, for with the increased size would go increased cost, great surplus of capacity, less frequent despatch, and lower efficiency generally. The unsuccessful Euston Tunnel of forty years ago—practically an underground railway—is an extreme illustration of this point, though in that case there were grave mechanical defects as well.

From a mechanical point of view the system has been brought to such perfection that it is no more experimental than a locomotive or an electric tramcar. The unique value of tube service is due to immediate despatch, high velocity of transit, immunity from traffic interruption, and economy. The greatest obstacle to rapid intercommunication is the delay resulting from accumulations due to time schedules. The function of tube service is to abolish time schedules and all consequent delays.

The number of trades parcels annually delivered in London is estimated at *more than 200,000,000*. A careful canvass has been made of 1,000 shops only, which represent a very small fraction of the total number in the county. As a result it has been ascertained that these 1,000 shops deliver no fewer than 60,000,000 parcels yearly, a fact that seems to more than justify the foregoing estimate; on the other hand, it is known from official data that the parcel post in London is represented by less than 25,000,000, or one-ninth of the total parcel traffic. With a tube system in operation, every parcel, instead of waiting for "the next delivery," would leave the shop immediately. After being despatched by the tube it would be delivered at a tube station within a quarter of a mile at least of its destination, and thence by messenger. The entire time consumed for an ordinary parcel would be not over an hour, and for a special parcel fifteen to twenty minutes. They require from three to six hours or longer at present.

The advantages of the tube system to the public would be manifold. Customers would find their purchases at home upon their return, or, if they preferred, could do their shopping by telephone, making their selections from goods sent on approval by tube. The shopman would find himself relieved from a vast amount of confusion and annoyance, less of his shop space given up to delivery, and his expenses reduced. Small shops would be able to draw upon wholesale houses for goods not in stock, while the customer waited. Such delay and confusion as are frequently occasioned by fogs would be reduced to a minimum.

While the success of the project is not dependent on Post Office support, the Post Office should be one of the greatest gainers by it. The time of delivery of local letters would be reduced from an average of three hours and six minutes to one hour. Express letters would be delivered more quickly than telegrams. This has been demonstrated conclusively again and again in New York and other American cities where the tubes have been in operation for years. The latest time of posting country letters would be deferred from one-half to one hour, and incoming letters would be advanced by a similar period. The parcels post would gain in precisely the same way, but to an even larger extent.

If the Post Office choose to avail themselves of the opportunity, every post office will become a tube station and every tube station a post office. Thus the same number of postmen covering but a tithe of the present distances could make deliveries without time schedules at intervals of a few minutes with a handful instead of a bagful of letters.

The sorting of mails would be performed at every station instead of at a few. Incoming country mails would be taken from the bags at the railway termini, and the same bags refilled with outgoing country mails, thus avoiding needless carriage to the Post Office and back. No bags at all would be used for local mails, the steel carriers themselves answering that purpose.

At every tube terminal a post-office clerk would be stationed, so that the mails would never for an instant be out of post-office control. Its absolute security would be further ensured by a system of locking, so that the carriers could only be opened by authorised persons at the station to which they were directed. These safeguards offer a striking contrast to the present method that entrusts mail bags to the sole custody of van drivers in the employ of private contractors.

If the mails were handled by tube, business men would be able to communicate with each other and receive replies several times in one day, and country and foreign letters could always be answered upon the day of receipt. The effect would be felt all over the Empire.

Would the laying of the tubes seriously impede traffic? The promoters assure us that the inconvenience would not be comparable to that caused by laying a gas, water, or telephone system. When one of those has been laid the annoyance, they urge, has only begun. The streets must be periodically reopened for the purpose of making thousands of house connections, extensions, and repairs. When a pneumatic tube is once down it is good for a generation at least. It is not subject to recurrent alterations incidental to house connections and repairs. In three American cities the tubes have been touched but three times in twelve years, and in those cases the causes were a bursting water main and faulty adjacent electric installations. The repairs were effected in a few hours.

From a general consideration of the scheme we may now turn to some mechanical details. The pipes would be of 1 foot internal diameter, made in 12-foot lengths. "Straight sections," writes an engineering correspondent of *The Times*, "would be of cast-iron, bored, counter-bored, and turned to a slight taper at one end, to fit a recess at the other end (of the next tube), to form the joints, which could be caulked. Joints made in this way are estimated to permit of a deflection of 2 inches from the straight, so that the laying and bedding need not be exact. Bent sections are to be of seamless brass; these are bored true before bending. The permissible curvature is determined upon the basis of a *maximum* bend of 1 foot radius for every 1 inch of diameter; the 1 foot diameter of the London tubes would consequently be allowed a *maximum* curvature of 12 foot radius. Measured at the enlarged end, the over-all diameter of each pipe is 17 inches, and as two such pipes are to be laid side by side, with 18 inches between centres, the clear width will be 35 inches. The trenches are therefore to be cut 36 inches wide, and in order to have a comparatively free run for the sections, it is proposed to cut the trenches 6 feet deep."

When the hundred miles of piping have been laid, the entire system will be tested to a pressure of 25 lbs. to the square inch, or about two and a half times the working pressure. Engines of 10,000 h.p. will be required to feed the lines with air, for the propulsion of the carriers, each 3 feet 10 inches long, and weighing 70 lbs.

In order to ensure the delivery of a carrier at its proper destination, whether a terminus or an intermediate station, Mr. Batcheller has made a most ingenious provision. On the front of a carrier is fixed a metal plate of a certain diameter. At each station two electric wires project into the tube, and as soon as a plate of sufficient diameter to short-circuit these wires arrives, the current operates delivery mechanism, and the carrier is switched off into the station box. The despatcher, knowing the exact size of disc for each station, can therefore make certain that the carrier shall not go astray.

It may occur to the reader that, should a carrier accidentally stick anywhere in the tubes, it would be a matter of great difficulty to locate it. Evidently one could not feel for it with a long rod in half a mile of tubing—the distance between every two stations—with much hope of finding it. But science has evolved a simple, and at the same time quite reliable, method of coping with the problem. M. Bontemps is the inventor. He located troubles in the Paris tubes by firing a pistol, and exactly measuring the time which elapsed between the report and its echo. As the rate of sound travel is definitely known, instruments of great delicacy enable the necessary calculations to be made with great accuracy. When a breakdown occurred on the Philadelphia tube line, Mr. Batcheller employed this method with great success, for a street excavation, made on the strength of rough measurements with the timing apparatus, came within a few feet of the actual break in the pipe, caused by a subsidence, while the carriers themselves were found almost exactly at the point where the workmen had been told to begin digging.[23]

There is no doubt that, were such a system as that proposed established, an enormous amount of time would be saved to the community. "A letter from Charing Cross to Liverpool Street," says *The World's Work*, "occupies by post three hours; by tube transit it would occupy twenty to forty minutes, or by an express system of tube transit ten to fifteen minutes. Express messages carried by the Post Office in London last year (1903) numbered about a million and a half, but the cost sometimes seems very heavy. To send a special message by hand from Hampstead to Fleet Street, for example, costs 1s. 3d., and takes about an hour. It is claimed that it could be sent by pneumatic tube at a cost of 3d. in from fifteen to twenty minutes, and that for local service the tube would be far quicker than the telegraph, and many times cheaper."

It has been calculated that from one-sixth to one-quarter of the wheeled traffic of London is occupied with the distribution of mails and parcels; and if the tubes relieved the streets to this extent, this fact alone would be a strong argument in their favour. It is impossible to believe that tube transmission on a gigantic scale will not come. Hitherto its development has been hindered by mechanical difficulties. But these have been mostly removed. In the United States, where the adage "time is money" is lived up to in a manner scarcely known on this side of the Atlantic, the device has been welcomed for public libraries, warehouses, railway depôts, factories—in short, for all purposes where the employment of human messengers means delay and uncertainty. Twenty years ago Berlier proposed to connect London and Paris by tubes of a diameter equal to that of the pipes contemplated in the scheme now before Parliament. Our descendants may see the tubes laid; for when once a system of transportation has been proved efficient on a large scale its development soon assumes huge proportions. And even the present generation may witness the tubes of our big cities lengthen their octopus arms till town and town are in direct communication. After all it is merely a question of "Will it pay?" We have the *means* of uniting Edinburgh and London by tube as effectually as by telephone or telegraph. And since the general trend of modern commerce is to bring the article to the customer rather than to give the customer the trouble of going to select the article *in situ*—this applies, of course, to small portable things only—"shopping from a distance" will come into greater favour, and the pneumatic tube will be recognised as a valuable ally. We can imagine that Mrs. Robinson of, say, Reading, will be glad to be spared the fatigue of a journey to Regent Street when a short conversation over the telephone wires is sufficient to bring to her door, within an hour, a selection of silver ware from which to choose a wedding present. And her husband, whose car has perhaps broken a rod at Newbury, will be equally glad of the quick delivery of a duplicate part from the makers. These are only two possible instances, which do not claim to be typical or particularly striking. If you sit down and consider what an immense amount of time and expense could be saved to you in the course of a year by a "lightning despatch," you will soon come to the conclusion that the pneumatic tube has a great future before it.

FOOTNOTE:

23. *Cassier's Magazine*, xiii, 456.

CHAPTER XXIII

AN ELECTRIC POSTAL SYSTEM

FAR swifter than the movements of air are those of the electric current, which travels many thousands of miles in a second of time.

Thirty miles an hour is the speed proposed for the pneumatic tube system mentioned in our last chapter. An Italian, Count Roberto Taeggi Piscicelli, has elaborated an electric post which, if realised, will make such a velocity as that seem very slow motion indeed.

Cable railways, for the transmission of minerals, are in very common use all over the world. At Hong-Kong and elsewhere they do good service for the transport of human beings. The car or truck is hauled along a stout steel cable, supported at intervals on strong poles of wood or metal, by an endless rope wound off and on to a steam-driven drum at one end of the line, or motion is imparted to it by a motor, which picks up current as it goes from the cable itself and other wires with which contact is made.

Count Piscicelli's electric post is an adaptation of the electric cableway to the needs of parcel and letter distribution.

At present the mail service between towns is entirely dependent on the railway for considerable distances, and on motors and horsed vehicles in cases where only a comparatively few miles intervene. London and Birmingham, to take an instance, are served by seven despatches each way every twenty-four hours. A letter sent from London in the morning would, under the most favourable conditions, not bring an answer the same day—at least, not during business hours. So that urgent correspondence must be conducted over either the telephone or the telegraph wires.

Count Piscicelli proposes a network of light cableways—four lines on a single set of supports—between the great towns of Britain. Each line—or rather track—consists of four wires, two above and two below, each pair on the same level. The upper

288

pair form the run-way for the two main wheels of the carrier; the lower pair are for the trailing wheels. Three of the wires supply the three-phase current which drives the carrier; the fourth operates the automatic switches installed every three or four miles for transforming the high-tension 5,000-volt current into low-tension 500-volt current in the section just being entered.

The carriers would be suitable for letters, book-parcels, and light packages. The speed at which they would move—150 miles per hour to begin with—would render possible a ten-minute service between, say, the towns already mentioned. The inventor has hopes of increasing the speed to 250 m.p.h., a velocity which would appear visionary had we not already before us the fact that an electric car, weighing many tons, has already been sent over the Berlin-Zossen Railway at $131\frac{1}{2}$ miles per hour. At any rate, the electric post can reasonably be expected to outstrip the ordinary express train. "Should such speeds as Count Piscicelli confidently discusses," says *The World's Work*, "be attained, they would undoubtedly confer immense benefits upon the mercantile and agricultural community—upon the agricultural community because in this system is to be found that avenue of transmission to big centres of population of the products of *la petite culture*, in which Mr. Rider Haggard, for example, in his invaluable book on *Rural England*, sees help for the farmer and for all connected with the cultivation of the soil. Count Piscicelli proposes to obviate the delays at despatching and receiving towns by an inter-urban postal system, in which the principal offices of any city would be connected with the head-office and with the principal railway termini. From each of the sub-offices would radiate further lines, along which post-collecting pillars are erected, and over which lighter motors and collecting boxes (similar to the despatch boxes) travel. The letter is put in through a slot and the stamp cancelled by an automatic apparatus with the name of the district, number of the post, and time of posting. The letter then falls into a box at the foot of the column. On the approach of a collecting-box the letter slot would be closed, and by means of an electric motor the receptacle containing the letters lifted to the top of the column and its contents deposited in the collecting-box, which travels alone past other post-collecting poles, taking from each its toll, and so on to the district office. Here, in a mercantile centre, a first sorting takes place, local letters being retained for distribution by postmen, and other boxes carry their respective loads to the different railway termini, or central office."

Were such an order of things established, there would be a good excuse for the old country woman who sat watching the telegraph wire for the passage of a pair of boots she was sending to her son in far away "Lunnon"!

CHAPTER XXIV

AGRICULTURAL MACHINERY

PLOUGHS — DRILLS AND SEEDERS — REAPING MACHINES — THRESHING MACHINES — PETROL-DRIVEN FIELD MACHINERY — ELECTRICAL FARMING MACHINERY

AGRICULTURE is at once the oldest and most important of all national industries. Man being a graminivorous animal—witness his molar, or grinding, "double" teeth—has, since the earliest times, been obliged to observe the seasons, planting his crops when the ground is moist, and reaping them when the weather is warm and dry. Apart from the nomad races of the deserts and steppes, who find their chief subsistence in the products of the date-palm and of their flocks and herds, all nations cultivate a large portion of the country which they inhabit. Ancient monuments, the oldest inscriptions and writings, bear witness to the prime importance of the plough and reaping-hook; and it may be reasonably assumed that the progress of civilisation is proved by the increased use of cereal foods, and better methods of garnering and preparing them.

For thousands of years the sickle, which Greek and Roman artists placed in the hand of their Goddess of the Harvest, and the rude plough, consisting of, perhaps, only a crooked bough with a pointed end, were practically the only implements known to the husbandman besides his spade and mattock. Where labour is abundant and each householder has time to cultivate the little plot which suffices for the maintenance of his own family, and while there is little inducement to take part in other than agricultural industries—tedious and time-wasting methods have held their own. But in highly civilised communities carrying on manufactures of all sorts it is difficult for the farmer to secure an abundance of human help, and yet it is recognised that a speedy preparation and sowing of the land, and a prompt gathering and threshing of the harvest, is all in favour of producing a successful and well-conditioned crop.

In England, eighty years ago, three men lived in the country for every one who lived in the town. Now the proportion has been reversed; and that not in the British Isles alone. The world does not mean to starve; but civilisation demands that as few people as possible should be devoted to procuring the "staff of life" for both man and beast.

We should reasonably expect, therefore, that the immense advance made in mechanical science during the last century should have left a deep mark on agricultural appliances. Such an expectation is more than justified; for are there not many among us who have seen the sickle and the flail at work where now the "self-binder" and threshing machine perform the same duties in a fraction of the time formerly required? The ploughman, plodding sturdily down the furrow behind his clever team, is indeed still a common sight; but in the tilling season do we not hear the snort of the steam-engine, as its steel rope tears a six-furrow plough through the mellow earth? When the harvest comes we realise even more clearly how largely machinery has supplanted man; while in the processes of separating the grain from its straw the human element plays an even smaller part. It would not be too much to say that, were we to revert next year to the practices of our grandfathers, we should starve in the year following.

This chapter will be confined to a consideration of machinery operated by horse, steam, or other power, which falls under four main headings,—ploughs, drills, reapers, and threshers.

PLOUGHS

The firm of Messrs. John Fowler and Company, of Leeds, is most intimately connected with the introduction of the steam plough and cultivator. Their first type of outfit included one engine only, the traversing of the plough across the field being effected by means of cables passing round a pulley on a low, four-wheeled truck, moved along the opposite edge of the field by ropes dragging on an anchor. Another method was to have the engine stationary at one corner of the field, and an anchor at each of the three other corners, the two at the ends of the furrow being moved for every journey of the plough. In, or about, the year 1865 this arrangement succumbed to the simple and, as it now seems to us, obvious improvement of introducing a second engine to progress vis-à-vis with the first, and do its share of the pulling. The modern eight-furrow steam plough will turn ten acres a day quite easily, at a much lower cost than that of horse labour. For tearing up land after a crop "cultivators" are sometimes used. They have arrowhead-shaped coulters, which cut very deep and bring large quantities of fresh earth to the surface.

The ground is now pulverised by harrows of various shapes, according to the nature of the crop to be sown. English farmers generally employ the spike harrow; but Yankee agriculturists make great use of the spring-tooth form, which may best be described as an arrangement of very strong springs much resembling in outline the springs of house bells. The shorter arm is attached to the frame, while the longer and pointed arm tears the earth.

DRILLS AND SEEDERS

In highly civilised countries the man carrying a basket from which he flings seeds broadcast is a very rare sight indeed. The primitive method may have been effective—a good sower could cover an acre evenly with half a pint of turnip seed—but very slow. We now use a long bin mounted on wheels, which revolves discs inside the bin, furnished with tiny spoons round the periphery to scoop small quantities of seed into tubes terminating in a coulter. The farmer is thus certain of having evenly planted and parallel rows of grain, which in the early spring, when the sprouting begins, make so pleasant an addition to the landscape.

The "corn," or maize, crop of the United States is so important that it demands special sowing machinery, which plants single grains at intervals of about eighteen inches. A somewhat similar device is used for planting potatoes.

Passing over the weeding machines, which offer no features of particular interest, we come to the

REAPING MACHINES,

on which a vast amount of ingenuity has been expended. At the beginning of the nineteenth century the Royal Agricultural Society of Great Britain offered a prize for the introduction of a really useful machine which should replace the scythe and sickle. Several machines were brought out, but they did not prove practical enough to attract much attention. Cyrus H. McCormick invented in 1831 the reaper, which, with very many improvements added, is to-day employed in all parts of the world. The most noticeable point of this machine was the bar furnished with a row of triangular blades which passed very rapidly to and fro through slots in an equal number of sharp steel points, against which they cut the grain. The to-and-fro action of the cutter-blade was produced by a connecting-rod working on a crank rotated by the wheels carrying the machine.

A WHEAT-CUTTER

A "heading reaper" being pushed over a wheat crop by six mules. It cuts off the ears only, leaving the straw standing. The largest machines of this type used in California take swathes 50 feet broad.

The first McCormick reaper did wonders on a Virginian farm; other inventors were stimulated; and in 1833 there appeared the Hussey reaper, built on somewhat similar lines. For twelve years or so these two machines competed against one another all over the United States; and then McCormick added a raker attachment, which, when sufficient grain had accumulated on the platform, enabled a second man on the machine to sweep it off to be tied up into a sheaf. At the Great Exhibition held in London in 1851, the judges awarded a special medal to the inventor, reporting that the whole expense of the Exhibition would have been well recouped if only the reaper were introduced into England. From France McCormick received the decoration of the Legion of Honour "for having done more for the cause of agriculture than any man then living."

It would be reasonable to expect that, after this public recognition, the mechanical reaper would have been immediately valued at its true worth. "Yet no man had more difficulty in introducing his machines than that pioneer inventor of agricultural implements. Farmers everywhere were slow to accept it, and manufacturers were unwilling to undertake its manufacture. Even after the value of the machine had been demonstrated, everyone seemed to fear that it would break down on rocky and uneven fields; and the inventor had to demonstrate in person to the farmers the practicability of the reapers, and then even guarantee them before the money could be obtained. Through all these trying discouragements the persistent inventor passed before he saw any reward for the work that he had spent half a lifetime in perfecting. The ultimate triumph of the inventor may be sufficient reward for his labours and discouragements, but those who would begrudge him the wealth that he subsequently made from his invention should consider some of the difficulties and obstacles he had to overcome in the beginning."[24]

In 1858 an attachment was fitted to replace the second passenger on the machine. Four men followed behind to tie up the grain as it was shot off the machine.

Inventors tried to abolish the need for these extra hands by means of a self-binding device.

A practical method, employing wire, appeared in 1860; but so great was the trouble caused by stray pieces of the wire getting into threshing and other machinery through which the grain subsequently passed that farmers went back to hand work, until the Appleby patent of 1873 replaced wire by twine. Words alone would convey little idea of how the corn is collected and encircled with twine; how the knot is tied by an ingenious shuttle mechanism; and how it is thrown out into a set of arms which collect sufficient sheaves to form a "stook" before it lets them fall. So we would advise our readers to take the next chance of examining a modern self-binder, and to persuade the man in charge to give as lucid an explanation as he can of the way in which things are done.

Popular prejudice having once been conquered, the success of the reapers was assured. The year 1870 saw 60,000 in use; by 1885 the output had increased to 250,000; and to-day the manufacture of agricultural labour-saving machines gives employment to over 200,000 people; an equal number being occupied in their transport and sale in all parts of the globe.

In California, perhaps more than in any other country, "power" agricultural machinery is seen at its best. Great traction-engines here take the place of human labour to an extraordinary extent. The largest, of 50 h.p. and upwards, "with driving-wheels 60 inches in diameter and flanges of generous width, travel over the uneven surface of the grain fields, crossing ditches and low places, and ascending the sides of steep hills, with as much apparent ease as a locomotive rolls along its steel rails. Such powerful traction-engines, or 'automobiles' as they are commonly called by the American farmers, are capable of dragging behind them sixteen 10-inch ploughs, four 6-foot harrows, and a drill and seeder. The land is thus ploughed, drilled, and seeded all at one time. From fifty to seventy-five acres of virgin soil can thus be ploughed and planted in a single day. When the harvest comes the engines are again brought into service, and the fields that would ordinarily defy the best efforts of an army of workmen are garnered quickly and easily. The giant harvester is hitched to the traction-engine in place of the ploughs and harrows, and cuts, binds, and stacks the golden wheat from seventy-five acres in a single day. The cutters are 26 feet wide, and they make a clear swathe across the field. Some of them thresh, clean, and sack the wheat as fast as it is cut and bound. Other traction-engines follow to gather up the sacked wheat, and whole train-loads of it thus move across the field to the granaries or railways of the seaboard or interior."

For "dead ripe" crops the "header" is often used in California. Instead of being pulled it is *pushed* by mules, and merely cuts off the heads, leaving the straw to be trampled down by the animals since it has no value. Swathes as wide as 50 feet are thus treated, the grain being threshed out while the machine moves.

One of the most beautiful, and at the same time useful, crops in the world is that of maize, which feeds not only vast numbers of human beings, but also countless flocks and herds, the latter eating the green stalks as well as the ripened grain. The United States alone produced no less than 2,523,648,312 bushels of this cereal in 1902, as against 987,000,000 bushels of wheat, and 670,000,000 bushels of barley. Now, maize has a very tough stalk, often 10 feet high and an inch thick, which cannot be cut with the ease of wheat or barley. So a special machine has been devised to handle it. The row of corn is picked up, if fallen, by chains furnished with projecting spikes working at an angle to the perpendicular, so as to lift and simultaneously pull back the stalks, which pass into a horizontal **V**-shaped frame. This has a broad opening in front, but narrows towards its rear end, where stationary sickles fixed on either side give the stalk a drawing cut before it reaches the single knife moving to right and left in the angle of the **V**, which severs the stalk completely. The McCormick machine gathers the corn in vertical bundles, and ties them up ready for the "shockers."

THRESHING MACHINES

In principle these are simple enough. The straw and grain is fed into a slot and pulled down between a toothed rotating drum and a fixed toothed concave. These tear out the grain from the ear. The former falls into the hopper of a winnowing and riddling machine, which clears it from dust and husks, and allows it to pass to a hopper. An endless chain of buckets carries it to the delivery bins, holding just one sackful each, which when full discharge the grain through spouts into the receptacles waiting below their mouths. An automatic counter records the number of sackfuls of corn that have been discharged, so that dishonesty on the part of employés becomes practically an impossibility. While the grain is thus treated, oscillating rakes have arranged the straw and shaken it out behind in a form convenient for binding, and the chaff has passed to its proper heap, to be used as fuel for the engine or as food for cattle.

PETROL-DRIVEN FIELD MACHINERY

On water, rail, and road the petrol engine has entered into rivalry with steam—very successfully too. And now it bids fair to challenge both steam-engine and horse as the motive power for agricultural operations. Probably the best-known English petrol-driven farmer's help is that made by Mr. Dan Albone, of Biggleswade, who in past times did much to introduce the safety bicycle to the public. The "Ivel" motor is not beautiful to look upon; its sides are slab, its outlines rather suggestive of an inverted punt. But it is a willing and powerful worker; requires no feeding in the early hours of the morning; no careful brush down after the day's work; no halts to ease wearied muscles. In one tank is petrol, in another lubricating oil, in a third water to keep the cylinders cool. A double-cylinder motor of 18 h.p. transmits its energy through a large clutch and train of cogs to the road wheels, made extra wide and well corrugated so that they shall not sink into soft ground or slip on hard. There is a broad pulley-wheel peeping out from one side of the machine, which is ready to drive chaff-cutters or threshers, pump, grind corn, or turn a dynamo at a moment's notice.

A MOTOR PLOUGH

The "Ivel" Agricultural Motor pulling a three-furrow plough. A motor thus harnessed will plough six acres a day at a total cost per acre of five shillings. It is also available for reaping, threshing, chaff-cutting, and other duties on a farm.

Hitch the "Ivel" on to a couple of reapers or a three-furrow plough, and it soon shows its superiority to "man's friend." Here are some records:—

Eleven acres, one rood, thirteen poles of wet loam land ploughed in $17\frac{1}{2}$ hours, at a cost per acre of 5s.

Nineteen acres of wheat reaped and bound in 10 hours, at a cost of 1s. 9d. per acre.

Fifteen acres, three roods of heavy grass cut in $3\frac{1}{2}$ hours, cost, 1s. per acre.

With horses the average cost of ploughing is about 10s. an acre; of reaping 5s. So that the motor does at least twice the work for the same money.

We may quote a paragraph from the pen of "Home Counties," a well-known and perspicacious writer on agricultural topics.

"It is because motor-farming is likely to result in a more thorough cultivation of the land and a more skilful and more enlightened practice of agriculture, and not in a further extension of those deplorable land-scratching and acre-grasping methods of which so many pitiful examples may be seen on our clay soils, that its beginnings are being sympathetically watched by many people who have the best interests of the rural districts and the prosperity of agriculture at heart."[25]

Will our farmers give the same welcome to the agricultural motor that was formerly accorded to the mechanical reaper? Prophecy is risky, but if, before a decade has elapsed, the horse has not been largely replaced by petrol on large farms and light land, the writer of these lines will be much surprised.

ELECTRICAL FARMING MACHINERY

In France, Germany, Austria, and the United States the electric motor has been turned to agricultural uses. Where water-power is available it is peculiarly suitable for stationary work, such as threshing, chaff-cutting, root-slicing, grinding, etc. The current can be easily distributed all over a large farm and harnessed to portable motors. Even ploughing has been done with electricity: the energy being derived either from a steam-engine placed near by, or from an overhead supply passing to the plough through trolley arms similar to those used on electric trams.

The great advances made recently in electrical power transmission, and in the efficiency of the electric motor, bring the day in sight when on large properties the fields will be girt about by cables and poles as permanent fixtures. All the usual agricultural operations of ploughing, drilling, and reaping will then be independent of horses, or of steam-engines panting laboriously on the headlands. In fact, the experiment has been tried with success in the United States. Whichever way we look, Giant Steam is bowing before a superior power.

FOOTNOTES:

24. *Cassier's Magazine.*

25. *The World's Work*, vol. iii. 499.

CHAPTER XXV

DAIRY MACHINERY

MILKING MACHINES — CREAM SEPARATORS — A MACHINE FOR DRYING MILK

MILKING MACHINES

THE farm labourer, perched on a three-legged stool, his head leaning against the soft flank of a cow as he squirts the milk in snowy jets into the frothing pail, is, like the blacksmith's forge throwing out its fiery spark-shower, one of those sights which from childhood up exercise a mild fascination over the onlooker. Possibly he or she may be an interested person in more senses than one, if the contents of the pail are ultimately to provide a refreshing drink, for milk never looks so tempting as when it carries its natural froth.

Modern methods of dairying demand the most scrupulous cleanliness in all processes. Pails, pans, and "churns" should be scoured until their shining surfaces suggest that on them the tiniest microbe could not find a footing. Buildings must be well aired, scrubbed, and treated occasionally with disinfectants. Even then danger may lurk unseen, and the milk is therefore for certain purposes sterilised by heating it to a temperature approaching boiling-point and simultaneously agitating it mechanically to prevent the formation of a scum on the surface. It is then poured into sealed bottles which bid defiance to exterior noxious germs.

The human hand, even if washed frequently, is a difficult thing to keep scientifically clean. The milkman has to put his hand now on the cow's side, now on his stool; in short, he is constantly touching surfaces which cannot be guaranteed germless. He may, therefore, infect the teats, which in turn infect the milk. So that, for health's sake as well as to minimise the labour and expense of milking, various devices have been tried for mechanically extracting the fluid from the udder. Many of these have died quick deaths, on account of their practical imperfections. But one, at least, may be pronounced a success—the Lawrence-Kennedy cow-milker, which is worked by electricity, and supplies another proof of the adaptability of the "mysterious fluid" to the service of man.

On the Isle de la Loge in the Seine is a dairy farm which is most up-to-date in its employment of labour-saving appliances, including that just mentioned. Here a turbine generates power to work vacuum pumps of large capacity. The pumps are connected to tubes terminating in cone-shaped rubber caps that can be easily slipped on to the teat; four caps branching out from a single suction chamber. As soon as they have been adjusted, the milkman —now shorn of a great part of his rights to that title—turns on the vacuum cock, and the pulsator, a device to imitate the periodic action of hand milking, commences to work. The number of pulsations per minute can be regulated to a nicety by adjusting screws. On its way to the pail the milk passes through a glass tube, so that the operator may see when the milking is completed.

This method eliminates the danger of hand contamination. It also protects the milk entirely from the air, and it has been stated that, when thus extracted, milk keeps sweet for a much longer time than under the old system. The cows apparently do not object to machinery replacing man, not even the Jersey breed, which are the most fidgety of all the tribe. Under the heading of economy the user scores heavily, for a single attendant can adjust and watch a number of mechanical milkers, whereas "one man, one cow" must be the rule where the hand is used. From the point of romance, the world may lose; the vacuum pump cannot vie with the pretty milkmaid of the songs. Practical people will, however, rest content with pure milk *minus* the beauty, in preference to milk *plus* the microbe and the milkmaid, who—especially when she is a man—is not always so very beautiful after all.

CREAM SEPARATORS

In the matter of separating the fatty from the watery elements of milk machinery also plays a part. The custom of allowing the cream to "rise" in open pans suffices for small dairies where speed and thoroughness of separation are not of primary importance. But when cream is required in wholesale quantities for the markets of large towns, or for conversion into butter, much greater expedition is needed.

The mechanical cream separator takes advantage of the laws of centrifugal force. Milk is poured into a bowl rotating at high speed on a vertical axis. The heavier—watery—portions climb up the sides of the bowl in their endeavour to get as far away as possible from the centre of motion; while the lighter particles of cream, not having so much momentum, are compelled to remain at the bottom. By a simple mechanical arrangement, the—very—skim milk is forced out of one tube, and the cream out of another. An efficient separator removes up to 99 per cent. of the butter fat. Small sizes, worked by hand, treat from 10 to 100 gallons of milk per hour; while the large machines, extensively used in "creameries," and turned by horse, steam, electric, or other power, have a capacity of 450 gallons per hour. The saving effected by mechanical methods of separation is so great that dairy-farmers can now make a good profit on butter which formerly scarcely covered out-of-pocket expenses incurred in its manufacture.

A MACHINE FOR DRYING MILK

Milk contains 87 per cent. of water and about 12 per cent. of nutritive matter. Milk which has had the water evaporated from it becomes a highly concentrated food, very valuable for many purposes which could not be served by the natural fluid. Until lately the process of separating the solid and liquid constituents was too costly to render the manufacture of "dried milk" a profitable industry. But now there is on the market a drying apparatus, manufactured by Messrs. James Milnes and Son, of Edinburgh, which almost instantaneously drives off the water.

The machine used for this—the Just-Hatmaker—process is simple. It consists of two large metal drums, 28 inches in diameter and 5 feet long, mounted horizontally in a framework with a space of about one-eighth of an inch between them. High-pressure steam, admitted to the drums through axial pipes, raises their surfaces to a temperature of 220° Fahr. The milk is allowed to flow in thin streams over the revolving drums, the heat of which quickly evaporates the water. A coating of solid matter gradually forms, and this is scraped off by a knife and falls into a receptacle.

The milk is not boiled nor chemically altered in any way, though completely sterilised by the heat. This machine promises to revolutionise the milk trade, as farmers will now be able to convert the very perishable product of their dairies into an easily handled and imperishable powder of great use for cooking and the manufacture of sweetmeats. Explorers and soldiers can have their milk supply reduced to tabloid form, and a pound tin of the lozenges will temper their tea or coffee over many a camp fire far removed from the domestic cow.

CHAPTER XXVI

SCULPTURING MACHINES

The savage who, with a flint point or bone splinter, laboriously scratched rude figures on the walls of his cave dwelling, did the best he was capable of to express the emotions which affect the splendidly equipped sculptor of to-day; he wished to record permanently some shape in which for the time he was interested, religiously or otherwise.

The sun, moon and stars figure largely in primitive religions as objects of worship. They could be easily suggested by a few strokes of a tool. But when mortals turned from celestial to terrestrial bodies, and to the worship of human or animal forms—the "graven images" of the Bible—a much higher level of art was reached by the sculptor, who endeavoured to give faithful representations in marble of the great men of the time and of the gods which his nation acknowledged.

The Egyptians, whose colossal monuments strew the banks of the Nile, worked in the most stubborn materials—basalt, porphyry and granite—which would turn the edge of highly tempered steel, and therefore raise wonder in our minds as to the nature of the tools which the subjects of the Pharaohs must have possessed. Only one chisel, of a bronze so soft that its edge turned at the first stroke against the rock under which it was found, has so far come to light. Of steel tools there is no trace, and we are left to the surmise that the ancients possessed some forgotten method of hardening other metals—including bronze—to a pitch quite unattainable to-day. Whatever were their implements, they did magnificent work; witness the splendid sculptures of vast proportions to be found in the British Museum; and the yet huger statues, such as those of Memnon and those at Karnak, which attract tourists yearly to Egypt.

The Egyptians admired magnitude; the Greeks perfection of outline. The human form in its most ideal development, so often found among a nation with whom athleticism was almost a religion, inspired many of the great classical sculptors, whose work never has been, and probably never will be, surpassed. Great honour awaited the winner in the Olympian games; but the most coveted prize of all was the permission given him—this after a succession of victories only—to erect a statue of himself in the sacred grove near the shrine of Olympian Jove. Happy the man who knew that succeeding generations would gaze upon a marble representation of some characteristic attitude assumed by him during his struggle for the laurel crown.

Until recently the methods of sculpture have remained practically unaltered for thousands of years. The artist first models his idea in clay or wax, on a small scale. He then, if he designs a life-size or colossal statue, erects a kind of iron skeleton to carry the clay of the full-sized model, copied proportionately from the smaller one. When this is finished, a piece-mould is formed from it by applying wet lumps of plaster of Paris all over the surface in such a manner that they can be removed piecemeal, and fitted together to form a complete mould. Into this liquid plaster is run, for a hollow cast of the whole figure, which is smoothed and given its finishing touches by the master hand.

This cast has next to be reproduced in marble. Both the cast and the block of marble are set up on "scale-stones," revolving on vertical pivots. An ingenious instrument, called a "pointing machine," now comes into play. It has two arms ending in fine metal points, movable in ball-and-socket joints. These arms are first applied to the model, the lower being adjusted to touch a mark on the scale-stone, the upper to just reach a mark on the figure. The operator then clamps the arms and revolves the machine towards the block of marble, the scale-stone of which has been marked similarly to its fellow. The bottom arm is now set to rest on the corresponding mark of the scale-stone; but the upper, which can slide back telescopically, is prevented from assuming its relative position by the unremoved portions of the block. The workman therefore merely notices the point on the block at which the needle is directed, and drills a hole into the marble on the line of the needle's axis, to a depth sufficient to allow the arm to be fully extended. This process is repeated, in some cases many thousands of times, until the block has been honeycombed with small holes. The carver can now strike off the superfluous marble, never going beyond the depth of a hole; and a rough outline of the statue appears. A more skilled workman follows him to shape the material to a close copy of the cast; and the sculptor himself adds the finishing touches which stamp his personality on the completed work.

Only a select few of the world's greatest sculptors have ventured to strike their statues direct from the marble, without recourse to a preliminary model. Such a one was Michelangelo, who, as though seized by a creative frenzy, would hew and hack a block so furiously that the chips flew off like a shower, continuing his attack for hours, yet never making the single false stroke that in the case of other masters has ruined the work of months. He truly was a genius, and must have possessed an almost supernatural faculty of knowing when he had reached the exact depth at any point in the great block of marble from which his design gradually emerged.

The formation of artistic *models* will always require the master's hand; but the *reproduction* of the cast in marble or stone can now be performed much more expeditiously than is possible with the pointing machine. We have already two successful mechanisms which in an almost incredibly short time will eat a statue out of a block in faithful obedience to the movement of a pointer over the surface of a finished design. They are the Wenzel Machine Sculptor and Signor Augusto Bontempi's *Meccaneglofo*.

THE WENZEL SCULPTURING MACHINE

In the basement of a large London business house we found, one dark November afternoon, two men at work with curious-looking frameworks, which they swayed backwards and forwards, up and down, to the accompaniment of a continuous clattering of metal upon stone. Approaching nearer, we saw, lying horizontally in the centre of the machine, a small marble statue, its feet clamped to a plate with deep notches in the circumference. On either side, at equal distances, were two horizontal blocks of marble similarly attached to similar plates. The workman had his eyes glued on a blunt-nosed pointer projecting from the middle of a balanced frame. This he passed slowly over the surface of the statue, and simultaneously two whirring drills also attached to the frame ate into the stone blocks just so far as the movement of the frame would permit. The drills were driven by electric power and made some thousands of revolutions per minute, throwing off the stone they bit away in the form of an exceedingly fine white dust.

It was most fascinating to watch the almost sentient performance of the drills. Just as a pencil in an artist's hands weaves line into line until they all suddenly spring into life and show their meaning, so did the drills chase apparently arbitrary grooves which united, spread, and finally revealed the rough-hewn limb.

Every now and then the machinist twisted the footplates round one notch, and snicked the retaining bolts into them. This exposed a fresh area of the statue and of the blocks to the pointer and the drills. The large, coarse drills used to clear away the superfluous material during the earlier stages of the work were replaced by finer points. The low relief was scooped out, the limbs moulded, the delicate curves of cheek and the pencilling of eyebrows and lips traced, and in a few hours the copies were ready for the usual smoothing and finishing at the hands of the human sculptor.

According to the capacity of the machine two, four, or six duplicates can be made at the cost of a little more power and time. Nor is it necessary to confine operations to stone and marble, for we were shown some admirable examples of wooden statues copied from a delicate little bronze, and, were special drills provided, the relations could be reversed, bronze becoming passive to motions controlled by a wooden original.

"Sculpturing made easy" would be a tempting legend to write over the Wenzel machine. But it would not represent the truth. After all, the mechanism only *copies*, it cannot originate, which is the function of the sculptor. It stands to sculpturing in the same relation as the printer's "process block" to the artist's original sketch, or the lithographic plates to the painter's coloured picture. Therefore prejudice against machine-made statues is as unreasonable as objection to the carefully-executed *replica* of a celebrated painting. The sculptor himself has not produced it at first hand, yet his personality has been stamped even on the copy, for the machine can do nothing except what has already been done for it. The machine merely displaces the old and imperfect "pointing" by hand, substituting a method which is cheaper, quicker, and more accurate in its interpretation of the model.

It is obvious that, apart from sculpture proper, the industrial arts afford a wide field for this invention. In architecture, for instance, carved wood and stonework for interiors and exteriors of buildings have been regarded hitherto as expensive luxuries, yet in spite of their cost they are increasingly indulged in. The architect now has at his disposal an economical method of carving which will enable him to utilise ornamental stonework to almost any degree. Sculptured friezes, cornices, and capitals, which, under the old régime, would represent months of highly paid hand labour, may now be reproduced rapidly and in any quantity by the machine, which could be adapted to work on the scaffolding itself.

What will become of the stonemasons? Won't they all be thrown out of work, or at least a large number of them? The best answer to these questions will be found in a consideration of industries in which machinery has replaced hand work. Has England, as a cotton-spinning nation, benefited because the power-loom was introduced? Does she employ more operatives than she would otherwise have done, and are these better paid than the old hand weavers? All these queries must have "Yes!" written against them. In like manner, if statuary and decoration becomes inexpensive, twenty people will be able to afford what hitherto was within the reach of but one; and an industry will arise beside which the output of the present-day monumental mason will appear very insignificant. The sculpturing machine undoubtedly brings us one step nearer the universal House Beautiful.

A complete list of the things which the versatile "Wenzel" can perform would be tediously long. Let it therefore suffice to mention boot-lasts, gun-stocks, moulds, engineering patterns, numeral letters, and other articles of irregular shape, as some of the more prosaic productions which grow under the buzzing metal points. Some readers may be glad to hear that the Wenzel promises another hobby for the individual who likes to "use his hands," since miniature machines are purchasable which treat subjects of a size not exceeding six inches in diameter. No previous knowledge of carving is necessary, and as soon as the elementary principles have been mastered the possessor of a small copier can take advantage of wet days to turn out statuettes, busts, and ornamental patterns for his own or friends' mantelpieces. And surely a carefully finished copy in white marble of some dainty classic figure or group will be a gift well worth receiving! The amateur photographer, the fret-sawyer, and the chip-carver will have to write "Ichabod" over their workshops!

The Wenzel has left its experimental stage far behind. The German Emperor, after watching the creation of a miniature bust of Beethoven, expressed his delight in a machine that could call a musician from lifeless stone. The whole of the interior decoration of the magnificent Rathaus, Charlottenburg, offers a splendid example of mechanical wood carving, which tourists would do well to inspect.

We may now pass to

THE BONTEMPI SCULPTURING MACHINE,

for such is the translation of the formidable word *Meccaneglofo*. This machine is the invention of Signor Augusto Bontempi, a native of Parma, who commenced life as a soldier in the Italian army, and while still young has won distinction as a clever engineer.

His machine differs in most constructional details from the Wenzel. To begin with, the pressure of the drills on the marble is imparted by water instead of by the hand; secondly, the block to be cut is arranged vertically instead of horizontally; thirdly, the index-pointer is not rigidly connected to the drill frame, but merely controls the valves of hydraulic mechanism which guides the drills in any required direction. The drills are *rotated* by electricity, but all their other movements come from the pressure of water.

A SMALL WENZEL AUTOMATIC SCULPTURING MACHINE

This cuts statuettes, two at a time, out of stone or wood, the cutters being guided by a pointer passed over the surface of the model by the girl.

Undoubtedly the most ingenious feature of the Bontempi apparatus is the pointer's hydraulic valve, which gives the drills a forward, lateral, or upward movement, or a compound of two or three movements. When the pointer is not touched all the valve orifices remain closed, and the machine ceases to work. Should the operator pull the pointer forwards a water-way is opened, and the liquid passes under great pressure to a cylinder which pushes the drill frame forward. If the pointer be also pressed sideways, a second channel opens and brings a second cylinder into action, and the frame as a whole is moved correspondingly, while an upward twist operates yet a third set of cylinders, and the workman himself rises with the drills.

As soon as the sensitive tip of the pointer touches an object it telescopes, and immediately closes the valves, so that the drills bore no further in that direction.

The original and copies are turned about from time to time on their bases in a manner similar to that already described in treating the Wenzel. As many as twenty copies can be made on the largest machines.

Quite recently there has been installed in Southwark, London, a gigantic Bontempi which stands 27 feet high, and handles blocks 5 feet 6 inches square by 10 feet high, and some 20 tons in weight. Owing to the huge masses to be worked only one copy can be made at a time; though, doubtless, if circumstances warranted the expense, a machine could be built to do double, triple, or quadruple duty. The proprietors have discovered an abrasive to grind granite—ordinary steel chisels would be useless—and they expect a great demand for columns and monumental work in this stubborn material, as their machines turn out finished stuff a dozen times faster than the mason.

An interesting story is told about the early days of Signor Bontempi's invention. When he set up his experimental machine at Florence, the workmen, following the example of the Luddites, rose in a body and threatened both him and his apparatus with destruction. The police had to be called in to protect the inventor, who thought it prudent to move his workshop to Naples, where the populace had broader-minded views. The Florentines are now sorry that they drove Signor Bontempi away, for they find that instead of depressing the labour market, the mechanical sculptor is a very good friend to both proprietor and employé.

NOTE.—For information and illustrations the author has to thank Mr. W. Hanson Boorne, of the Machine Sculpture Company, Aldermary House, London, E.C., and Mr. E. W. Gaz, secretary of the Automatic Sculpture Syndicate, Sumner Street, Southwark.

CHAPTER XXVII

AN AUTOMATIC RIFLE

While science works ceaselessly to cure the ills that human flesh is heir to, invention as persistently devises weapons for man's destruction. Yesterday it was the discoveries of Pasteur and the Maxim gun; to-day it is the Finsen rays and the Rexer automatic rifle.

Though one cannot restrain a sigh on examining a new contrivance, the sole function of which is to deal out death and desolation—sadly wondering why such ingenuity might not have been directed to the perfecting of a machine which would render life more easy and more pleasant; yet from a book which deals with modern mechanisms we may not entirely exclude reference to a class of engines on which man has expended so much thought ever since gunpowder first entered the arena of human strife.

We therefore choose as our subject for this chapter a weapon hailing from Denmark, a country which, though small in area, contains many inventors of no mean repute.

In a London office, within sight of the monument raised to England's great sailor hero, the writer first made acquaintance with the Rexer gun, which, venomous device that it is, can spit forth death 300 times a minute, though it weighs only about 18 lbs.

Its form is that of an ordinary rifle of somewhat clumsy build. The eye at once picks out a pair of supports which project from a ring encircling it near the muzzle. Even a strong man would find 18 lbs. too much to hold to his shoulder for any length of time; so the Rexer is primarily intended for stationary work. The user lies prone, rests the muzzle on its supports, presses the butt to his shoulder, and blazes away. History repeats itself in the chronicles of firearms, though it is a very long way from the old matchlock supported on a forked stick to the latest thing in rifles propped up by two steel legs.

Machine-guns, such as the Maxim and Hotchkiss, weigh 60 lbs.

and upwards, and have to be carried on a wheeled carriage, drawn either by horses or by a number of men. In very rough country they must be loaded on pack-horses or mules. When required for action, the gun, its supports and appliances, separated for packing, must be hurriedly reassembled. This means loss of valuable time.

The Rexer rifle can be carried almost as easily as a Lee-Metford or Mauser, and fires the ordinary small-bore ammunition. Wherever infantry or cavalry can go, it can go too, without entailing any appreciable amount of extra haulage.

Before dealing with its actual use as a fighting arm we will notice the leading features of its construction.

The gun comprises the stock, the casing and trigger-plate which enclose the breech mechanism, the barrel, and the perforated barrel cover, to which are attached the forked legs on which the muzzle end is supported when firing, and which fold up under the cover when not in use. The power for working the mechanism is obtained from the recoil, which, when the gun is fired, drives the barrel, together with the breech and the other moving parts, some two inches backwards, thus compressing the powerful recoil-spring which lies behind the breech, enclosed in the front part of the stock, and which, after the force of the recoil is spent, expands, and thus drives the barrel forward again into the firing position. The recoil and return of the breech operate a set of levers and other working parts within the casing, which, by their combined actions following one another in fixed order, open the breech, eject the empty cartridge-case, insert a new cartridge into the chamber, and close the breech; and when the gun is set for automatic action, and the gunner keeps his finger pressed on the trigger, the percussion arm strikes the hammer and the cartridge is fired; the round of operations repeating itself till the magazine is emptied, or until the gunner releases the trigger and thereby interrupts the firing.

A noticeable feature is the steel tube surrounding the barrel. It is pierced with a number of openings to permit a circulation of air to cool the barrel, which is furnished with fins similar to those on the cylinder of an air-cooled petrol motor to help dissipate the heat caused by the frequent explosions. Near the ends of the cover are the guides, in which the barrel moves backwards and forwards under the influence of the recoil and the recoil-spring. The supports are attached to the casing in such a way that the stock of the gun can be elevated or depressed and traversed through considerable angles without altering the position of the supports on the ground. The rear end of the barrel cover is firmly fixed to the casing of the breech mechanism, and forms with this and the stock the rigid part of the gun in which the moving portions work, their motions being guided and controlled by cams and studs working in grooves and notches and on blocks attached to the rigid parts.

Without the aid of special diagrams it is rather hard to explain the working of even a simple mechanism; but the writer hopes that the following verbal description, for which he has to thank the Rexer Company, will at least go some way towards elucidating the action of the breech components.

Inside the casing is the breech, the front end of which is attached rigidly to the barrel, the rear end being in contact with the recoil arm, which is directly operated by the recoil spring lying in a recess in the stock. In the breech is the breech-block, which has three functions: first to guide the new cartridges from the distributer, which passes them from the magazine one by one into the casing, to the firing position in the chamber (*i.e.* the expanded part of the bore at the rear end of the barrel); secondly, to hold the cartridge firmly fixed in the chamber, and to act as an abutment or support to the back of the cartridge when it is fired, and thus transmit the backward force of the explosion to the recoil spring; thirdly, to allow the spent cartridges to be discharged from the chamber by the extractor, and to direct them by means of a guide curved downwards from the chamber, so that they may be flung through an opening provided for that purpose in the trigger-plate in front of the trigger, and out of the way of the gunner. (This opening is closed by a cover when the gun is not in use, and opens automatically before the shot can be fired.) In order to effect this threefold object, the breech-block is pivoted in the rear to the rear of the breech, and has a vertical angular motion within it, so that the fore end of the block can move into three different positions in relation to the chamber: one, below the chamber to guide the cartridge into it; one, directly in line with the chamber, to back the cartridge; and one, above the chamber, to allow the ejection of the spent cartridge-case by the extractor. The cartridge is fired by a long pin through the breech-block, struck behind by a hammer operated by a special spring.

The first function of the breech-block is, as we have said, to act as a guide for the cartridge into the chamber ready for firing, after the fashion of the old Martini-Henry breech-block. The actual pushing forward of the cartridge is performed by a lever sliding on the top of the block. After the explosion a small vertical lever jerks out the cartridge-case against the block, and causes it to cannon downwards through the aperture in the trigger-plate already mentioned.

On the left-hand side of the breech casing is a small chamber, open at the top and on the side next the breech. To the top is clipped the magazine, filled with twenty-five cartridges. The magazine is shaped somewhat like a slice of melon, only that the curved back and front are parallel. The sides converge towards the inner edge. It is closed at the lower end by a spring secured by a catch. When a magazine is attached to the open top of the chamber the catch is released so as to put chamber and magazine in direct communication. The cartridges would then be able to drop straight into the breech chamber through the side slot, were the latter not protected by a curved horizontal shutter, called the distributer. Its action is such that when a cartridge is being passed through into the breech casing, the shutter closes, and holds the remaining cartridges in the magazine; and when the cartridge has passed it opens and lets the next into position in the side casing.

As soon as a cartridge enters the breech it is pushed forward into the chamber ready for firing by the feeder lever. The magazine and the holder are so arranged that when the last cartridge has passed from the magazine to the distributer, the motion of the moving parts of the gun is arrested till the magazine is removed, when the motion is resumed so far as to push the remaining cartridge into the chamber and bring the breech-block into the firing position. When another magazine has been fixed in the holder, firing can be resumed by pulling the trigger; but if another magazine is not fixed in the holder the last cartridge cannot be fired by pulling the trigger, and only by pulling a handle which will be presently described. This arrangement secures the continuance of the automatic firing being interrupted only by the very brief interval required for charging the apparatus.

The gun is fired, as usual, by pulling a trigger. If a steady pull be kept on the trigger the whole contents of the magazine will be fired automatically (the last cartridge excepted); but if such continuous firing is not desired, a few shots at a time may be fired automatically by alternately pulling and releasing the trigger. If it is desired to fire shot by shot from the magazine, a small swivel on the trigger-guard is moved so as to limit the movement of the trigger. By moving this swivel out of the way, automatic firing is resumed. The gun may also be fired without a magazine by simply feeding cartridges by hand into the magazine holder. In front of the trigger-guard is a safety catch, and if this is set to "safe" the gun cannot be fired until the catch is moved to "fire."

It is obvious that the recoil cannot come into action until a shot has been fired. A handle is therefore provided on the right-hand side outside the casing, by means of which the bolt forming the axis of the recoil and percussion arms may be turned so as to imitate the action of the recoil. This handle must be turned to bring the first cartridge into the chamber, but this having been done, the handle returns to its normal position, and need not be moved again.

We may now watch a gunner at work. He chooses his position, opens out the supports, and pushes them into the ground so as to give the muzzle end a firm bearing. He then takes a magazine from the box he carries with him, and fixes it by a rapid motion into the magazine holder, then, resting his left hand on the stock to steady it, he pulls over the handle with his right so as to bring the barrel and all the moving mechanism into the backward position. He then releases the handle, and the recoil spring comes into action and drives the breech forward, when the controlling gear brings the front end of the breech-block into its downward position, admits the first cartridge into the breech and pushes it forward by the cartridge-feeder into the barrel chamber. The breech-block then rises to its central position at the back of the cartridge, and the gun is ready for firing.

If automatic firing is required, the gunner sets the swivel at the back of the trigger in the right position, sights the object at which he has to fire, and pulls the trigger, thereby exploding the first cartridge. The recoil then drives back the barrel and the breech. The breech-block is moved into its highest position, making room for the ejection of the empty cartridge-case, which is then ejected by the extractor. At the end of the recoil the block falls into its lowest position, the cartridge-feeder having then arrived at the back of the breech-block. The recoil-spring now drives the breech forward, admits the new cartridge on to the breech-block and drives it forward by the feeder into the chamber. The breech-block rises to its position behind the cartridge and is locked in that position. The percussion arm is then released automatically, strikes the hammer, and fires the second cartridge, the cycle of operations repeating itself till the last cartridge but one has been fired, when the magazine is charged and the cycle of operations is again renewed and continued till the second set of cartridges has been fired. The operations follow one another with such rapidity that the twenty-five cartridges contained in the magazine can be fired in less than two seconds. At the same time, the rate of firing remains under the control of the gunner, who can interrupt it at any moment by simply releasing the trigger. He can also alter his aim at any time and keep it directed on a moving object and fire at any suitable moment.

THE "REXER" AUTOMATIC MACHINE GUN

It only weighs $17\frac{1}{2}$ lb., and can fire 300 shots per minute. The crescent-shaped clips hold 25 cartridges each, and as soon as one has been emptied another can be affixed in a moment.

In service it is not intended that every man should be armed with a Rexer, but only 3 to 5 per cent., constituting a separate detachment which would act independently of the artillery and other machine-guns. The latter would, as at present, cover the infantry's advance up to within some 500 yards of the enemy, but at this point would have to cease firing for fear of hitting their own men. This period, when the artillery can neither shoot over the heads of their infantry, nor bring up the guns for fear of losing the teams, affords the golden opportunity for the Rexer, which is advanced with the firing line. If the fire of the detachment were concentrated on a part of the enemy's line, that portion would be unable to reply while the attacking force rushed up to close quarters. One hundred men armed with Rexers would be as valuable as several hundred carrying the ordinary service weapon, while they would be much more easily disposed, advanced, or withdrawn.

A squadron of cavalry would be accompanied by three troopers armed with Rexers and by one leading a pack-horse laden with extra magazines. Each gunner would have on his horse

400 cartridges, and the pack-horse 2,400 rounds, distributed in leather cases over a specially designed saddle. When a squadron, not provided with machine-guns, has to open a heavy fire, a considerable proportion must remain behind the firing line to hold the horses of the firing party. When, on the other hand, Rexers are present, only a few men would dismount, leaving the main body ready to charge at the opportune moment; and, should the attack fail, they could cover the retreat.

A use will also be found for the Rexer in fortresses and on war vessels; in fact, everywhere where the machine-gun can take a part.

After exhaustive trials, the Danish Government has adopted this weapon for both army and navy; and it doubtless will presently be included in the armament of other governments. There are signs that the most deadly arm of the future will be the automatic rifle. Perhaps a pattern even lighter than the Rexer may appear. If every unit of a large force could fire 300 rounds a minute, and ammunition were plentiful, we could hardly imagine an assault in which the attacking party would not be wiped out, even if similarly armed; for with the perfection of firearms the man behind cover gets an ever-increasing advantage over his adversary advancing across the open.

A BALL-BEARING RIFLE

Rapidity of fire is only one of the desirable features in a firearm. Its range—or perhaps we had better say its muzzle velocity—is of almost equal importance. The greater this is, the flatter is the trajectory or curve described by the bullet, and the more extended the "point blank" range and the "danger zone."

Take the case of two rifles capable of flinging a bullet one mile and two miles respectively. Riflemen seldom fire at objects further off than, say, 1,200 yards; so that you might think that, given correct sighting in the weapon and a positive knowledge of the range, both rifles would have equal chances of making a hit.

This is not the fact, however, for the more powerful rifle sends its bullet on a course much more nearly parallel to the ground than does the other. Therefore an object six feet high would evidently run greater risks of being hit *somewhere* by the two-mile rifle than by the one-mile. Thus, if at 1,200 yards the bullet had fallen to within six feet of the ground, it might not actually strike earth till it had travelled 1,400 yards; whereas with a lesser velocity and higher curve, the point of impact might be only fifty yards behind. Evidently a six-foot man would be in danger anywhere in a belt 200 yards broad were the high-velocity rifle in operation, though the danger zone with the other weapon would be contracted to fifty yards.

At close quarters a flat trajectory is even more valuable, since it diminishes the need for altering the sights. If a rifle's point-blank range is up to 600 yards, you can fire at a man's head anywhere within that distance with a good chance of hitting him. The farther he is away, the lower he will be hit. A high trajectory would necessitate an alteration of the sights for every fifty yards beyond, say, two hundred.

The velocity of a projectile is increased—(1) by increasing the weight of the driving charge; (2) by decreasing the friction between the barrel and the projectile.

An American inventor, Mr. Orlan C. Cullen, has adopted a means already well tried in mechanical engineering to decrease friction.

He has produced a rifle, the barrel of which has in its walls eight spiral grooves of almost circular section, a small arc of the circle being cut away so as to put the groove in continuous communication with the bore of the barrel. These grooves are filled with steel balls, one-tenth of an inch in diameter, which are a good fit, and on the slot side of the groove project a very tiny distance into the barrel. The bullet—of hard steel—as it is driven through the barrel does not come into contact with the walls, but runs over the balls, which grip it with sufficient force to give it a spinning motion. The inventor claims that there is no appreciable escape of gas round the bullet, as the space between it and the barrel is so minute.

The ball races, or grooves, extend back to the powder chamber and forward to the muzzle. Their twist ceases a short distance from the muzzle to permit the insertion of recoil cushions, which break the forces of the balls as they are dragged forward by the bullet.

Mr. Cullen holds that a rifle built on this principle gives 40 per cent. greater velocity than one with fixed rifling—to be exact, has a point-blank range of 650 yards as compared with 480 yards of the Lee-Metford, and will penetrate 116 planks 1 inch thick each.

The absence of friction brings absence of heat, which in the case of machine-guns has always proved a difficulty. It also minimises the recoil, and reduces the weight of mountings for large guns.

Whether these advantages sufficiently outweigh the disadvantages of complication and cleaning difficulties to render the weapon acceptable to military authorities remains to be seen. We can only say that, if the ball bearing proves as valuable in ballistics as it has in machinery, then its adoption for firearms can be only a matter of time.

PLYMOUTH: W. BRENDON AND SON, LTD., PRINTERS.

www.ingramcontent.com/pod-product-compliance
Lightning Source LLC
Chambersburg PA
CBHW071411180526
45170CB00001B/58